AF746053

Stochastic Optimal Control and the U.S. Financial Debt Crisis

Jerome L. Stein

Stochastic Optimal Control and the U.S. Financial Debt Crisis

Springer

Jerome L. Stein
Division of Applied Mathematics
Brown University Box F
Providence, RI, 02912, USA

ISBN 978-1-4614-3078-0 ISBN 978-1-4614-3079-7 (eBook)
DOI 10.1007/978-1-4614-3079-7
Springer New York Heidelberg Dordrecht London

Library of Congress Control Number: 2012932090

© Springer Science+Business Media New York 2012
This work is subject to copyright. All rights are reserved by the Publisher, whether the whole or part of the material is concerned, specifically the rights of translation, reprinting, reuse of illustrations, recitation, broadcasting, reproduction on microfilms or in any other physical way, and transmission or information storage and retrieval, electronic adaptation, computer software, or by similar or dissimilar methodology now known or hereafter developed. Exempted from this legal reservation are brief excerpts in connection with reviews or scholarly analysis or material supplied specifically for the purpose of being entered and executed on a computer system, for exclusive use by the purchaser of the work. Duplication of this publication or parts thereof is permitted only under the provisions of the Copyright Law of the Publisher's location, in its current version, and permission for use must always be obtained from Springer. Permissions for use may be obtained through RightsLink at the Copyright Clearance Center. Violations are liable to prosecution under the respective Copyright Law.
The use of general descriptive names, registered names, trademarks, service marks, etc. in this publication does not imply, even in the absence of a specific statement, that such names are exempt from the relevant protective laws and regulations and therefore free for general use.
While the advice and information in this book are believed to be true and accurate at the date of publication, neither the authors nor the editors nor the publisher can accept any legal responsibility for any errors or omissions that may be made. The publisher makes no warranty, express or implied, with respect to the material contained herein.

Printed on acid-free paper

Springer is part of Springer Science+Business Media (www.springer.com)

In memory of
DAVID MORTON STEIN 1988–2008

Contents

List of Figures

List of Tables

Chapter 1
Introduction

Abstract The theme of this book is that the application of Stochastic Optimal Control (SOC) is very helpful in understanding and predicting debt crises. The mathematical analysis is applied empirically to the financial debt crisis of 2008, the crises of the 1980s and concludes with an analysis of the European debt crisis. I use SOC to derive a theoretically founded quantitative measure of an optimal, and an excessive leverage/debt/risk that increases the probability of a crisis. The optimal leverage balances risk against expected growth. The environment is stochastic: the capital gain, productivity of capital and interest rate are stochastic variables, and for an insurance company, such as AIG, the claims are also stochastic. I associate the housing price bubble with the growth of household debt. A bubble is dangerous insofar as it induces a non-sustainable debt. This danger is exacerbated insofar as a complex financial system is based upon it.

The Financial Crisis Inquiry Commission (FCIC) was created to examine the causes of the financial and economic crisis in the US. It asked: How did it come to pass that in 2008 our nation was forced to choose between two stark and painful alternatives – either risk the total collapse of our financial system and economy or inject trillions of taxpayer dollars into the financial system?

While the vulnerabilities that created the potential for crisis were years in the making, the collapse of the housing bubble – fueled by low interest rates and available credit, scant regulation and toxic mortgages – was the spark that ignited a string of events, that led to a full-blown crisis in the fall of 2008. Trillions of dollars of risky mortgages had become embedded throughout the financial system, as mortgage related securities were packaged, repackaged, and sold to investors around the world. When the bubble burst, hundreds of billions of dollars in losses in mortgages and mortgage related securities shook markets and financial institutions that had significant exposures to those mortgages and had borrowed heavily against them. This happened, not just in the US but around the world.

J.L. Stein, *Stochastic Optimal Control and the U.S. Financial Debt Crisis*,
DOI 10.1007/978-1-4614-3079-7_1, © Springer Science+Business Media New York 2012

Mortgage originators such as Countrywide sell packages of mortgages, household debt to the major banks. The latter in turn structure the packages and tranche them into senior, mezzanine and equity tranches. The income from the mortgages then flows like a waterfall. The senior tranche has the first claim, the mezzanine has the next and the equity tranche gets what, if anything is left. The illusion was that this procedure diversified risk and that relatively riskless tranches could be constructed from a mélange of mortgages of dubious quality.

The securities firms finance the purchases from short term loans from banks and money market funds, either repos secured by mortgages or commercial paper. The securities firms then sell the collateralized debt obligations CDOs, the mezzanine and equity tranches as packages to international investors, investment banks such as Merrill Lynch, Citi-group, Goldman-Sachs and hedge funds. These purchasers finance the purchases by short term bank borrowing. Securities firms and hedge funds may buy Credit Default Swaps (CDS) from companies such as AIG as insurance against declines in the values of the CDOs. If the mortgagors are unable to service their debts – the income from the mortgages declines – the repercussions are felt all along the line. This is a systemic risk that was ignored.

Despite the post crisis expressed view of many on Wall St. and in Washington that the crisis could not have been foreseen or avoided, the FCIC argued there were warning signs. The tragedy was that Washington and Wall St. ignored the flow of toxic mortgages and could have set prudent mortgage-lending standards. The Federal Reserve was the one entity empowered to do so and did not.

Regulators had ample power to protect the financial system and they chose not to use it. SEC could have required more capital and halted risky practices at the big investment banks. It did not. The Federal Reserve Bank of N.Y. (FRNY) and other regulators could have clamped down on Citigroup's excesses in the run up to the crisis. They did not. The dramatic failures of corporate governance and risk management at many systemically important financial institutions were a key cause of this crisis.

Many financial institutions as well as too many households borrowed to the hilt, leaving them vulnerable to financial distress or ruin if the value of their investments declined even moderately. As of 2007 the five major investment banks – Bear Stearns, Goldman Sachs, Lehman Brothers, Merrill Lynch and Morgan Stanley were operating with thin layers of capital – leverage ratios as high as 40:1. Less than a 3% drop in asset values would wipe out the firm.

A key institution in the financial crisis was AIG. At its peak it was one of the largest and most successful companies in the world. AIG's senior management ignored the terms and risks of the company's $79 billion derivatives exposure to mortgage related securities. The financial crisis put its credit rating under pressure, because AIG lacked the liquidity to meet collateral demands. In a matter of months AIG's worldwide empire collapsed.

The government was ill prepared for the crisis and its inconsistent response added to the uncertainty and panic in financial markets. It had no comprehensive and strategic plan for containment, because it lacked a full understanding of the risks and interconnection in the financial markets.

Prior to the crisis, it appeared to the academic world, financial institutions, investors, and regulators alike that risk had been conquered. The capital asset pricing model (CAPM) developed by Markowitz, Sharpe and Lintner explained the pricing of securities and how to manage risk. The options pricing model of Black, Scholes and Merton was used to construct financial derivatives with desired risk-expected returns combinations. Using these techniques, physicists, mathematicians and computer scientists – *the Quants* – were attracted to Wall St. to use good mathematics to manufacture financial derivatives.

Investors held highly rated securities they thought were sure to perform; the banks thought that they had taken the riskiest loans off their books; and regulators saw firms making profits and borrowing costs reduced. But each step in the mortgage securitization pipeline depended upon the next step to keep demand going.

The Fed and the IMF, who employed large numbers of PhD's in economics, were charged with surveillance of financial markets. The Fund surveillance reports reflect the state of the art – the quality of the models – in the economics profession. There was no fear of a financial crisis because the prevailing view was that they were the consequences of monetary excesses. The pre crisis period was the Great Moderation: moderate money growth and inflation and satisfactory real growth. Hence no cause to worry.

The Independent Evaluation Office (IEO) of the IMF assessed the performance of the IMF surveillance in the run up to the global financial crisis. It found that the IMF provided few clear warnings about the risks and vulnerabilities associated with the impending crisis before its outbreak in the US and elsewhere. For example, in spite of the fact that Iceland's banking sector had grown from about 100% of GDP in 2003 to almost 1,000% in 2007, the Fund did not recognize that this was a vulnerability that needed to be addressed urgently. Just before the crisis the IMF wrote that Iceland's medium term prospects remained enviable. They did not consider that Iceland's high leverage posed a risk to the financial system. The banner message was one of continued optimism after more than a decade of benign economic conditions and low macroeconomic volatility.

The IMF and the economics profession missed key elements that underlay the developing crisis. There was a "group think" mentality: this homogeneous group of economists in the Fund only considered issues within the prevailing paradigm in economics and there were no significant challenges to this point of view. The key assumption was that market discipline and self-regulation would be sufficient to stave off serious problems in financial institutions.

Neither the Fed nor the IMF discussed, until the crisis had already erupted, the deteriorating lending standards for mortgage financing, or adequately assessed the risks and impact of a major housing price correction on financial institutions. In fact the IMF praised the US for its light touch regulation and supervision that ultimately contributed to the problems of the financial system. Moreover, the IMF recommended that other advanced countries follow the US/UK approach. The Fund did not see the similarities between developments in the US and UK and the experience of other advanced economies and emerging markets that had previously faced financial crises.

1.1 The Subject and Contributions of This Book

The Dodd-Frank (D-F) bill establishes the Financial Services Oversight Council. The bill authorizes the Federal Reserve Board to act as agent for the Council to monitor the financial services marketplace to identify potential threats to the stability of the US financial system and to identify global trends and developments that could pose systemic risks to the stability of the US economy and to other economies. Neither the Fed nor the IMF, who based their analysis upon the dominant economic paradigm, has demonstrated its ability to fulfill these requirements. The techniques used by the Quants and rating agencies, based upon the dominant stochastic models, proved inadequate.

The four major studies of the US financial crisis are: Greenspan's *Retrospective* (2010); the Financial Crisis Inquiry Commission *Report* (FCIC 2011); Congressional Oversight Panel (COP 2010) *The AIG Rescue*, Its Impact on Markets and the Government's Exit Strategy; Congressional Oversight Panel (COP 2009), *Special Report on Regulatory Reform.* There is a large economics literature on the crisis in conference volumes and journals. They cover the same ground as the four major studies above and are primarily descriptive. Several discuss regulation and capital requirements but their recommendations are not based upon an optimizing framework. They do not provide analytical tools to answer the questions: (Q1) What is a theoretically founded quantitative measure of an optimal leverage? (Q2) What is an excessive risk that increases the probability of a crisis? (Q3) What is the explanatory power of the analysis?

The *theme of this book* is that the application of Stochastic Optimal Control is very helpful in understanding and predicting debt crises and in evaluating risk management. I associate the housing price bubble with the growth of household debt. *A bubble is dangerous insofar as it induces a non-sustainable debt.* This danger is exacerbated insofar as a complex financial system is based upon it. My analysis uses Stochastic Optimal Control (SOC) to answer questions (Q1)–(Q3) above. The optimal capital requirement/leverage balances risk against expected growth. The environment is stochastic: the capital gain, productivity of capital and interest rate are stochastic variables, and for an insurance company, such as AIG, the claims are also stochastic. In this manner the SOC approach developed in this book satisfies the requirements of the D-F bill described above.

There is a large economics literature that describes the crisis. There is a large mathematics literature on stochastic optimal control. My book synthesizes the two approaches. It is aimed at economists and mathematicians who are interested in understanding how SOC based techniques could have been useful in providing early warning signals of the recent crises, and at those interested in risk management. Key issues below are the subjects of the subsequent chapters and constitute the theme and contribution of this book.

Chapter 2 explains why the financial markets, and the Fed/IMF/economics profession, failed to anticipate the mortgage/housing and financial crisis and the vulnerability of AIG. They used inappropriate models and hence incorrectly evaluated risk and the probability of bankruptcy/ruin. The crucial ultimate variable is the household debt, the mortgage debt. The rest of the financial system rested upon the ability of the mortgagors to service their debts. *Systemic risk* describes the effects of the failure of the mortgagors to service their debts upon the financial structure. The leverage of the financial system transmitted the housing market shock into a collapse of the financial system.

A bubble is in effect a large positive "excess, unsustainable debt". Detection of a bubble corresponds to the detection of an "excess debt". The aim of this book is to derive an optimal debt/net worth ratio and excess debt ratio. The latter is equal to the difference between the actual and the optimal debt. The fundamentals are reflected in the optimal debt. The housing price bubble, its subsequent collapse, and the financial crisis were not predicted by either the market, the Fed, the IMF or regulators in the years leading to the crisis. Moreover, the Fed and Treasury rejected the warnings based upon publicly available information, and successfully advocated deregulation of Over The Counter (OTC) markets. As a result, transparency of prices was reduced, risk was concentrated in a few major financial institutions, and high leverage was induced. These were basic ingredients for the subsequent crisis.

The Fed, the IMF and Treasury lacked adequate tools, which might have indicated that asset values were vastly out of line with fundamentals. The Fed and the Fund were not searching for such tools because they did not believe that they could or should look for misaligned asset values or excess debt, despite warnings from Shiller, some people in the financial industry, the GAO, state bank regulators and FDIC. The Fed was blind-sided by the Efficient Market Hypothesis (EMH), that current prices reveal all publicly available information. One cannot second – guess the market. There cannot be an ex-ante misalignment. Bubbles exist only in retrospect. The Jackson Hole Consensus gave them great comfort in adopting a hands off position by claiming that "As long as money and credit remain broadly controlled, the scope for financing unsustainable runs in asset prices should also remain limited.. . .numerous empirical studies have shown that almost all asset price bubbles have been accompanied, if not preceded by strong growth of credit and or money". Since the period preceding the crisis was the Great Moderation, there was no need to worry.

So it was not just a lack of appropriate tools that undid the Fed; it was a complete lack of appreciation of what its role should be in heading off an economic catastrophe. There are two separate but related questions: Are identification and containment of a financial bubble legitimate activity of the Fed, and if they are, what are the best tools to carry out this analysis.

Former chairman of the Federal Reserve Board Alan Greenspan has great knowledge of financial markets. I think that his behavior may be explained rationally. First he understands that the function of financial markets is to channel saving into investment in the optimal way to promote growth. Second, like most of the economics profession, he or his staff accepted the generality of the *First Theorem of Welfare Economics*. This theorem states that a Competitive Equilibrium is a Pareto Optimum. The implication is that "market regulation" is superior to regulation by bureaucrats, politicians. Do not try to second guess the markets.

The belief in the generality of the *First Theorem of Welfare Economics* may have provided a basis for Greenspan's position. The Theorem does not hold in financial markets for several reasons. First, financial assets are not arguments in the utility function of households so that it makes little sense to say that the relative asset prices equal marginal rates of substitution. There is no tangency of indifference curves with the price line. Second, the assumption of atomistic agents operating in perfectly competitive markets with full information and stable preferences is wildly unrealistic. The Efficient Market Hypothesis EMH was a major foundation of Greenspan's view and that of the finance profession.

Chapter 3 considers the role of the "Quants"/mathematical finance. They are the physicists, mathematicians and computer scientists who were attracted to Wall St. The mathematics per se was not at fault in the crisis, but the finance models used were inadequate and grossly underestimated risk.

The finance literature was based upon the Efficient Market Hypothesis (EMH), the Black-Scholes-Merton (BSM) options price model and the CAPM. The EMH claims that asset markets are, to a good approximation, informationally efficient. Market prices contain most information about fundamental value. Prices of traded assets already reflect all publicly available information. The CAPM provides a good measure of risk. Assets can only earn high average returns if they have high betas. Average returns are driven by beta because beta reflects the extent that the addition of a small quantity of the asset to a diversified portfolio adds to the volatility of the portfolio. On the basis of the EMH and CAPM, Greenspan, the Fed and the finance profession believed that markets would be self-regulating through the activities of analysts and investors. Government intervention weakens the more effective private regulation.

Securitization/tranching, the CDOs and derivatives of derivatives produced an environment where the EMH/CAPM lost relevance. These bundles of many mortgage based securities seemed to tailor risk for different investors. Securitization/tranching gave the illusion that one could practically eliminate risk from risky assets and led to very high leverage. Ratings of the tranches were not based upon the quality of the underlying mortgages. They were all in the same bundle. The rating depended upon who got paid first in the stack of loans. The key question was how to rate and price the tranches. The issue concerned the correlation of the tranches. If a pool of loans started experiencing difficulties, and a certain percent of them defaulted, what would be the impact upon each tranche? The "apples in the basket model" assumed that they were like apples in a basket with a certain fraction of them being rotten. If one apple is rotten, it says nothing about whether the next apple chosen is rotten. Another very different one is "the slice of bread in the loaf" model. In that model if a slice (tranche) of bread is moldy, what is the probability that the next slice – or the rest of the loaf – is moldy? The Quants falsely assumed independence of tranches and assumed that they could tranche packages of "toxic assets" to produce a riskless tranche.

The Quants ignored how the interactions of the firms affected the return on the CDOs. The collapse of one group led to severe losses in groups before and after it in the chain. For example, the collapse of AIG affected the prices of "safe" as well as of risky assets. They based their estimates of risk upon the recent non-sustainable distribution of housing prices. They ignored the "no free lunch" constraint that capital gains cannot consistently exceed the mean interest rate. Most important, they ignored publicly available information concerning systemic risk. Their models ignored the systemic risk that the mortgagors would be unable to repay debt. The prices of many of the securities traded were opaque and estimated using arbitrary computer models. Hence the values of assets and liabilities on balance were not reflective of what they could fetch if sold.

Chapter 4 discusses the philosophy of the stochastic optimal control (SOC) techniques used in later Chaps. 5, 6 and 7. Modeling is crucial in economics and finance. Fisher Black, who developed the equation for options modeling, argued that given the models' limitations, "the right way to engage with a model is, like a fiction reader or a really great pretender, is to suspend disbelief and push it as far as possible... But then, when you've done modeling, you must remind yourself that ... although God's world can be divined by principles, humanity prefers to remain mysterious. Catastrophes strike when people allow theories to take on a life of their own and hubris evolves into idolatry."(quoted in Derman).

The net worth of the real estate sector in Chap. 5, and of AIG on Chap. 6, evolve dynamically. In the first case, debt is incurred in period t to purchase assets whose return is uncertain, and must be repaid in period t + 1 at an uncertain interest rate. In the second case, insurance is sold in period t and the claims in period t + 1 are uncertain. What is the optimal debt in the first case and what are the optimal insurance liabilities in the second case?

I discuss the strengths and limitations of alternative criterion functions, what should the firm or industry maximize? How should risk aversion be taken into account? Then I discuss the modeling of reasonable stochastic processes of the uncertain variables. Given the criterion function, each stochastic process implies a different quantitative, but similar qualitative, optimum debt/net worth or insurance liabilities/net worth. Using SOC I derive quantitative measures of an optimal and an excessive leverage, an excessive risk that increases the probability of a crisis. The optimal capital requirement or leverage balances risk against expected growth and return. The implications of the analysis are described graphically in the text and proved mathematically in an appendix. As the actual debt ratio exceeds the optimal ratio the expected growth declines and the risk rises. Thereby the probability of a debt crisis is directly related to the excess debt, the actual less optimal. A bubble is an unsustainable excess debt. The second part of the chapter discusses the models used in the insurance, or actuarial literature, concerning the probability of ruin. They are then compared with the SOC approach.

Chapter 5 applies this SOC analysis to the US financial crisis. I discuss the importance of the housing/real estate sector to the financial sector, and the characteristics of the mortgage market. Then two models of the stochastic process on the capital gain and interest rate are presented. Each implies a different value of the optimal debt/net worth. In order to do an empirical analysis, I derive an upper bound of the optimal debt ratio, based upon the alternative models, to derive a measure the excess debt: actual less the upper bound of the optimal ratio. The derived excess debt is shown to be an early warning signal (EWS) of the debt crisis as early as 2004.

Finally, the shadow banking system is discussed. The financial crisis was precipitated by the mortgage crisis for several reasons. First, a whole structure of financial derivatives was based upon the ultimate debtors – the mortgagors. Insofar as the mortgagors were unable to service their debts, the values of the derivatives fell. Second, the financial intermediaries whose assets and liabilities were based upon the value of derivatives were very highly leveraged. Changes in the values of their net worth were large multiples of changes in asset values. Third, the financial intermediaries were closely linked – the assets of one group were liabilities of another. A cascade was precipitated by the mortgage defaults. Since the "Quants" were following the same rules, the markets could not be liquid. In this manner, the mortgage debt crisis turned into a financial crisis.

Chapter 6 concerns insurance, the AIG case. First, I describe what happened to AIG in the 2007–2008 crisis. Then I evaluate the actuarial literature on optimal risk and capital requirements for insurers – Cramér-Lundeberg, ruin problems. I explain how SOC is a much more powerful tool of analysis. The stochastic optimal (SOC) approach's components are: the criterion function, the stochastic differential equations, and the stochastic processes. The solution for the optimal insurance liability/claims requirement on the basis of SOC follows. The chapter concludes with an evaluation of the government bailout.

AIG seriously underestimated risk because it ignored the negative correlation between the capital gain on insured assets and the liabilities/claims on AIG. The CDS claims grew when the value of the insured obligations CDO declined. This set off collateral requirements, and the stability of AIG was undermined. The chapter concludes with an evaluation of the government bailout.

Chapter 7 concerns the agricultural crisis of the 1980s and the S&L crisis in the 1980s. I explain that these crises had many features in common, but were localized. The crisis of 2007–2008 shared the common elements of the earlier two but was more pervasive and severe due to the financial structure that was based upon the housing/mortgage sector. This focus is upon the crisis of the 1980s, in particular the agriculture crisis. The policy issues are: How should creditors, banks and bank regulators evaluate and monitor risk of an excessive debt that significantly increases the probability of default? I show how the same techniques of stochastic optimal control used in Chaps. 5 and 6 are useful in providing early warning signals for the agricultural crisis. In the concluding part I compare the S&L crisis to the agricultural crisis.

Chapter 8 goes beyond the US financial crisis of 2008 and explains the inter country differences in the debt crisis in Europe. This subject is timely and I cannot ignore it. The external debts of the European countries are at the core of the current European crises. Generally, the crises are attributed to government budget deficits in excess of the values stated in the Stability and Growth Pact (SGP)/Maastricht treaty.

Proposals for reform generally involve increasing the powers of the European Union to monitor fiscal policies of the national governments and increasing bank regulation.

I explain: (a) to what extent the crises in the different countries were due to government budget deficits/government dissaving or to the private investment less private saving, (b) what is the mechanism whereby the actions of the public and private sectors lead to an unsustainable debt burden, defined as the ratio of debt service/GDP. The Stability and Growth Pact/Maastricht Treaty and the European Union focused upon rules concerning government debt ratios and deficit ratios. They ignored the problem of "excessive" external debt ratios in the entire economy that led to a crisis in the financial markets.

The techniques of analysis in this chapter differ from those in the previous chapters. In the previous chapters the debt ratio was a control variable. Using stochastic optimal control, I derived optimal debt ratios. This is *normative* economics. Chapter 8 is concerned with *positive* economics. The external debt ratio is not a control variable, but is an endogenous variable that is determined by "fundamentals" in a dynamic manner. The "fundamentals" are determined by the actions of both the public and the private sectors. I explain this by drawing upon the Natural Real Exchange Rate NATREX model (Stein 2006) of the equilibrium real exchange rate and external debt – the endogenous variables.

In this book, I do not discuss policy issues: regulation and reform. A Dissenting Statement by Wallison, in the Financial Crisis Inquiry Commission Report is: "The question that I have been most frequently asked about the Financial Crisis Inquiry Commission [FCIC] is why Congress bothered to authorize it all. Without waiting for the Commission's insights into the causes of the financial crisis, Congress passed and the President signed the Dodd-Frank Act (DFA), [with] far reaching and highly consequential regulatory legislation." The focus of my book is positive economics, and I avoid the political, normative, divisive and sociological aspects that regulation entails.

The history of this book reflects my debts to many people. When I retired from the economics department I was invited in 1997 by the Division of Applied Mathematics (DAM) to be a visiting professor/research. I had worked with Ettore Infante (DAM) for a decade in the 1960–1970s applying deterministic optimal control to economic problems in feedback form. I felt that I was returning home. Wendell Fleming and I started to discuss how and to what extent the techniques of stochastic optimal control can be useful in economics. Wendell is renowned for his contributions to pure and applied mathematics, and his book with Ray Rishel is essential reading. We decided that the debt crises would be an appropriate subject of interdisciplinary research. Thus I had to learn the mathematics literature using dynamic programming to determine what is an optimal trajectory of the debt. Our regular meetings resulted in our first article, Fleming and Stein (2004) in the Journal of Banking and Finance. I was invited to give a paper at the AMS-IMS-SIAM Research Conference in Mathematical Finance (2003), where I explained how one can successfully apply the techniques of the H-J-B equation to the crises of the 1980s. This was my first contact with the elite in the profession. They were masters of the techniques but were unaware of what one could do with them for real world problems in economics. I edited and contributed to a special issue of Australian Economic Papers "Stochastic Models in Economics and Finance" (2005). I was then invited to give a paper at a mathematics conference at the University of Wisconsin/Milwaukee applying the mathematical techniques to the US balance of payments. There I got to know Ray Rishel, who has been most helpful to me. I then edited and contributed to a Special Issue of the Journal of Banking and Finance "Intertemporal Optimization in a Stochastic Environment" (2007).

EUROPT invited me twice, once to Prague and once to Lithuania, to give keynote addresses about different aspects of my work. I was the only economist on the programs, the rest were mathematicians and O/R experts.

The next phase consisted of writing a series of articles aimed at economists under the rubric "Greenspan, Dodd-Frank and Stochastic Optimal Control". My aim was to explain how the failures of the Fed could have been avoided had the Fed used my techniques.

I felt that I had done all that I could to bring my work to the attention of the various professions. However, the Springer-Science editor Brian Foster wrote that there are many books on stochastic control and many descriptive books on the crisis, but none applied the techniques of SOC to the crises. Would I consider doing a book on the subject? Springer published Fleming-Rishel, so my book would be a nice complement.

It was unclear to me who could be the readership? I consulted Seth Stein, the author of Disaster Deferred on earthquake prediction. His advice was to write the book that I want to write and not write it with any specific constituency in mind. He suggested how I should present the mathematics in a way that both mathematicians and economists would benefit. He has been a constant source of excellent advice.

I had the good fortune to receive the advice and criticism from several sources. Peter Clark, Serge Rey, Karlhans Sauernheimer, Christoph Fischer and Carl D'Adda have been my economics critics. They have suggested many changes in points of view. Wendell Fleming and Ray Rishel have been my mathematics critics. Ren Cheng (Fidelity Investments) and Robert Selvaggio (Rutter Associates) have my consultants on what has been going on in the finance industry.

References

Congressional Oversight Panel, COP. 2009. "Special Report on Regulatory Reform". Affairs, New York.

Congressional Oversight Panel. 2010. *The AIG rescue, its impact on markets and the government's exit strategy*. Washington, DC: U.S. G.P.O.

Derman, Emanuel. 2004. *My life as a quant*. New York: Wiley.

Financial Crisis Inquiry Report. 2011. FCIC, Public Affairs, N.Y.

Fleming, Wendell H., and Jerome L. Stein. 2004. Stochastic optimal control, international finance and debt. *Journal of Banking and Finance* 28: 979–996.

Greenspan, Alan. 2010. The Crisis, Brookings papers on Economic Activity.

International Monetary Fund, Independent Evaluation Office. 2011. *IMF performance in the run-up to the financial and economic crisis*. Washington, DC: International Monetary Fund.

Stein, Jerome L. 2006. *Stochastic optimal control, international finance and debt crises*. New York: Oxford University Press.

Stein, Seth. 2010. *Disaster deferred*. New York: Columbia University Press.

Chapter 2
The Fed, IMF and Disregarded Warnings

Abstract The Fed, the IMF and Treasury lacked adequate tools, which might have indicated that asset values were vastly out of line with fundamentals. The Fed and the Fund were not searching for such tools because they did not believe that they could or should look for misaligned asset values or excess debt, The Fed was blind-sided by the Efficient Market Hypothesis (EMH), that current prices reveal all publicly available information. One cannot second- guess the market. There cannot be an ex-ante misalignment. Bubbles exist only in retrospect. The Jackson Hole Consensus gave them great comfort in adopting a hands off position by claiming that "As long as money and credit remain broadly controlled, the scope for financing unsustainable runs in asset prices should also remain limited.. ..numerous empirical studies have shown that almost all asset price bubbles have been accompanied, if not preceded by strong growth of credit and or money". Since the period preceding the crisis was the Great Moderation, there was no need to worry.

The theme of this chapter is to explain why the financial markets, the Fed, IMF, and economics profession failed to anticipate the mortgage housing and financial crisis. *I associate a bubble, such as in housing prices or agricultural prices, with an "excess debt*". There were debt crises in the 1980s – the agricultural crisis and the S&L crisis. They were very similar to the 2007–2008 financial crisis. The big difference was that the agricultural crisis was localized. On the other hand the housing sector and financial sectors were highly interrelated and leveraged. The Fed lacked a theoretical model with explanatory power to evaluate *systemic* risk and the probability of bankruptcy/ruin resulting from debt. The Fed and then Chairman Greenspan did not understand how to measure what is an "excessive" debt or leverage or unduly low capital requirement that will raise the probability of a crisis.

This chapter is organized as follows. First I discuss the Fed's and Greenspan's views. Second, I discuss the market anticipations, disregarded warnings and why the financial market failed to anticipate the crisis. Third, I discuss the controversy over deregulation. Fourth, I discuss the failures of the IMF and economics profession.

J.L. Stein, *Stochastic Optimal Control and the U.S. Financial Debt Crisis*,
DOI 10.1007/978-1-4614-3079-7_2, © Springer Science+Business Media New York 2012

2.1 Greenspan's Theme and the Fed

Prior to the subprime crisis of 2007, there was a false sense of safety in financial markets. Alan Greenspan (2004a) said "...the surge in mortgage refinancings likely improved rather than worsened the financial condition of the average homeowner". Moreover "Overall, the household sector seems to be in good shape, and much of the apparent increase in the household sector's debt ratios in the past decade reflects factors that do not suggest increasing household financial stress".

The market and the Fed did not consider these mortgages to be very risky. In February 2004, a few months before the Fed formally ended a run of interest rate cuts, Greenspan (2004b) said that "...improvements in lending practices driven by information technology have enabled lenders to reach out to households with previously unrecognized borrowing capacity. This extension of lending has increased overall household debt but has probably not meaningfully increased the number of households with already overextended debt." By 2007, a measure of risk, the yield spread (CCC bonds – 10 year US Treasury) fell to a record low.

Fed Chairman Ben Bernanke said (2005) in his testimony before Congress's Joint Economic Committee that US house prices have risen by nearly 25% over the past 2 years. However, these increases "largely reflect strong economic fundamentals" such as strong growth in jobs, incomes and the number of new households".

The failure to realize that there was an unsustainable bubble that would damage the world economy was pervasive. As late as April 2007, the IMF noted that "...global economic risks *declined* since...September 2006...The overall US economy is holding up well ...[and] the signs elsewhere are very encouraging." The venerated credit rating agencies bestowed credit ratings that implied AAA smooth sailing for many a highly toxic derivative product.

In 2008 Greenspan said "Those of us who have looked to the self-interest of lending institutions to protect stockholders' equity, myself included, are in a state of disbelief". In his retrospective he asks: could the breakdown have been prevented? The Fed was lulled into complacency about a bursting of the bubble and its aftermath because of recent history. First, they anticipated that the decline in home prices would be gradual. Second, there were only modestly negative effects of the 1987 stock market crash. The injections of Fed liquidity apparently helped stabilize the economy.

Greenspan's paper (2010) presents his retrospective view of the crisis. His theme has several parts. First, the decline and convergence of world real long term interest rates – not Federal Reserve monetary policy – led to significant housing price appreciation, a housing price bubble. This bubble was leveraged by debt. There was a heavy securitization of subprime mortgages. In the years leading to the current crisis, financial intermediation tried to function on too thin layer of capital – high leverage – owing to a misreading of the degree of risk embodied in ever more complex financial products and markets. Second, when the bubble unraveled, the leveraging set off a series of defaults. Third, the breakdown of the bubble was unpredictable and inevitable, given the "excessive" leverage – or unduly low capital – of the financial intermediaries. Fourth, the lesson for the future is that is imperative that there be an increase in regulatory capital and liquidity requirements by banks.

2.1.1 The Jackson Hole Consensus

Otmar Issing (former chief economist for the European Central bank, ECB) discussed the Lessons to be learned by Central banks from the recent financial crisis. The main thrust of his argument was a criticism of the Jackson Hole Consensus (JHC 2005) for the relation between asset price bubbles and the conduct of monetary policy.

During the boom years, abundant liquidity and low interest rates led to a situation of excessive risk taking and asset price bubbles. The JHC has been the prevailing regulatory approach taken by the Fed. It is based upon three principles. Central banks: (1) should not target asset prices, (2) should not try to prick an asset price bubble, (3) should follow a "mopping up" strategy after the bubble bursts by injecting enough liquidity to avoid serious effects upon the real economy. A justification for this policy was seen in the period 2000–2002 with the collapse of the dot.com bubble. The "mopping up" seemed to work well and there were no serious effects upon the real economy from following the JHC.

Issing objects to the JHC because it constitutes an asymmetric approach. When asset prices rise without inflationary effects measured by the CPI, this is deemed irrelevant for monetary policy. But when the bubble bursts, central banks must come to the rescue. This, he argues, produces a moral hazard. He notes that although the JHC strategy worked well in the 2000–2002 period it should not have justified the assumption that it would work afterwards in other cases. The JHC strategy certainly did not work in the 2007–2008 crisis that was precipitated by the bursting of the housing price bubble. He wrote: "Did we really need a crisis that brought the world to the brink of a financial meltdown to learn that the philosophy which was at the time seen as state of the art was in fact dangerously flawed?...we must conduct a thorough discussion as to appropriate strategy of central banks with respect to asset prices."

Issing favors giving the central banks a mandate for macro-prudential supervision. The ECB should be responsible for identifying macroeconomic imbalances and for issuing warnings and recommendations addressed to national policy makers. The "solution" proposed is one that monitors closely monetary and credit developments as the potential driving forces for consumer price inflation in the medium to short run. "As long as money and credit remain broadly controlled, the scope for financing unsustainable runs in asset prices should also remain limited." He notes: "numerous empirical studies have shown that almost all asset price bubbles have been accompanied, if not preceded by strong growth of credit and or money".

Table 2.1 Money, CPI, house price index, real GDP, percent change from previous year

Year	Narrow money	CPI	Real GDP	House price
1997	7.4% pa	2.3	4.5	2.8
1998	11.6	1.5	4.4	5.22
1999	12.4	2.2	4.8	4.44
2000	8.1	3.4	4.1	6.2
2001	15.7	2.8	1.1	8.1
2002	12.8	1.6	1.8	6.5
2003	7.3	2.3	2.5	7.1
2004	3.8	2.7	3.6	8.21
2005	2.1	3.4	3.1	12.7
2006	4.3	3.2	2.7	12.2

Source: Col. 1–3, Federal Reserve Bank St. Louis; Col. 4, Office of Federal Housing Price Oversight

However, these studies such as reported by the BIS are vague, inconclusive and not helpful. Even their authors conclude that the existing literature provides little insight into the recent financial crisis. The key question that is of concern to central banks and supervisory authorities is: When should credit growth be judged "too fast"? Moreover, contrary to Issing and BIS, it is very difficult to find a relation between recent money growth and the 2007–2008 financial crisis.

Table 2.1 contains the growth rates of narrow money, CPI inflation and house prices from the previous year. It is clear that in the years 2004–2006 leading up to the crisis of 2008, money growth and inflation were moderate but the inflation of house prices – the asset bubble – was high. Issing's "solution" does not have relevance for the recent crisis. High rates of growth of money and credit are sufficient, but not necessary, conditions for a financial crisis.

The Jackson Hole Consensus explains to a considerable extent the Fed's behavior. Greenspan has great knowledge of financial markets and did have some qualms about the housing boom. I think that his behavior can be explained rationally. First he understands that the function of financial markets is to channel saving into investment in the optimal way to promote growth. Second, like most of the economics profession, he or his staff accepted the generality of the *First Theorem of Welfare Economics*. This theorem (Koopmans and Bausch) states that: a Competitive Equilibrium is a Pareto Optimum. A Competitive Equilibrium is a vector of prices, where (1) supply equals demand, (2) consumers optimize demand and their supply of labor services, given their preferences and (3) producers optimize by maximizing their profits, given the technology. A Pareto Optimum is a vector of choices such that (4) supply equals demand and (5) it is not possible to select vectors which would make some people better off without making others worse off. The implication is that "market regulation" is superior to regulation by bureaucrats or politicians. Do not try to second guess the markets.

The belief in the generality of the *First Theorem of Welfare Economics* may have provided a basis for Greenspan's position. The Theorem does not hold in financial market for several reasons. First, financial assets are not arguments in the utility function of households so that it makes little sense to say that the relative asset prices equal marginal rates of substitution. There is no tangency of indifference curves with the price line. Second, the assumption of atomistic agents operating in perfectly competitive markets with full information and stable preferences is wildly unrealistic. The Efficient Market Hypothesis EMH was a major foundation of Greenspan's view and that of the finance profession. This hypothesis and its use by the Quants and beta as a measure of risk is discussed in Chap. 3.

2.1.2 Desirable Leverage, Capital Requirements

When the crash occurred, Greenspan wrote (2008) "Those of us who have looked to the self-interest of lending institutions to protect stockholders' equity, myself included, are in a state of disbelief". It is now widely believed that "excessive" leveraging, or an "excessive" debt ratio, at key financial institutions helped convert the initial subprime turmoil in 2007 into a full blown financial crisis of 2008. The ratio of debt L(t)/net worth X(t) is the debt ratio, and is denoted f(t) = L(t)/X(t) . Leverage is the ratio of assets/net worth A(t)/X(t) and is equal to one plus the debt ratio. Although leverage is a valuable financial tool, "excessive" leverage poses a significant risk to the financial system. For an institution that is highly leveraged, changes in asset values highly magnify changes in net worth. To maintain the same debt ratio when asset values fall either the institution must raise more capital or it must liquidate assets.

In his Retrospective Greenspan has qualified his unquestioned faith in the financial markets to allocate saving optimally to investment. The question is what should be done to rectify the problem? Regulation per se cannot be an improvement. Regulators are inclined to raise capital requirements to lower risk without considering expected return. He argues that there are limits to the level of regulatory capital if resources are to be allocated efficiently. A bank or financial intermediary requires significant leverage if it is to be competitive. Without adequate leverage, markets do not provide a sufficiently high rate of return on financial assets to attract capital to that activity. Yet, at too great a degree of leverage, bank solvency is at risk. The crucial question is what is a "desirable" degree of leverage? Since this is a main question of concern, I present Greenspan's views that I shall relate later in Chaps. 5, 6 and 7 to my Stochastic Optimal Control (SOC) analysis.

Greenspan suggests that the focus be on desirable capital requirements for banks and financial intermediaries. He starts with an identity, Equation (i) or (ii) for the rate of return on net worth r(t). This is income/equity. Net worth and equity are used interchangeably here. I will use my notation instead of his for the sake of consistency. Leverage is assets/equity = A(t)/X(t) or capital requirement is X(t)/A(t). Net income is Y(t). Define net income/assets Y(t)/A(t) = b(t).

(i) net income/equity = (net income/assets)(assets/equity).
(ii) rate of return on equity $r(t) = b(t)\ A(t)/X(t)$

He observes, that over the long run, there has been a remarkable stability in the ratio of net income/equity. It has ranged around 5% pa. Call this long run value r. Greenspan considers the long run ratio r without a time index as a required rate of return to induce the US banking system to provide the financial sector with the resources to promote growth. Equation (iii) must be satisfied. The minimum rate of return at any time r(t) should be equal to the long run value r.

(iii) $\min r(t) = b(t)A(t)/X(t) = r.$

Alternatively the maximum capital requirement X(t)/A(t) should satisfy (iv) or the minimum leverage should satisfy (v). If the capital requirement exceeds b(t)/r then – given the return on assets b(t) – the return on net worth falls below the required rate r.

(iv) $\max X(t)/A(t) = b(t)/r.$
(v) $\min A(t)/X(t) = r/b(t).$

Given the estimate r = 0.05, and the ratio b(t) of income/assets in the years prior to the crisis b(t) = 0.012, the maximum capital requirement should satisfy (vi) or minimum leverage should satisfy (vii).

(vi) $\max X(t)/A(t) = b(t)/r = 0.012/0.05 = 0.24.$
(vii) $\min A(t)/X(t) = 0.05/0.012 = 4.17.$

The maximum capital requirement X(t)/A(t) is 0.24, or minimum leverage is 4.17. A capital requirement greater than 0.24 depresses the rate of return r(t) below the required rate r.

Greenspan's derivation of desirable leverage has several advantages but leaves open several questions. First, the advantage of (vi) and (vii) is that it is an attempt to find a capital requirement or leverage that is sufficient to attract capital into the financial system. Second, it is a time varying ratio that takes into account b(t) the return on assets. However, risk is not explicit in his formulation. There is no explicit trade off between growth and risk. Third, the required minimum return on equity r is arbitrary and lacks theoretical foundations. Fourth, it says nothing about the effects upon risk and growth of leverage or capital requirements that deviate from the value in (vi) and (vii). The SOC analysis attempts to rectify these difficulties where the objective is to find a debt ratio, leverage or capital requirement that optimally balances expected growth against risk. The context is that the future is unpredictable, stochastic.

2.1.3 Market Anticipations of the Housing: Mortgage Debt Crisis

The subprime market was the trigger for the crisis because the income from the highly leveraged financial intermediaries was ultimately based upon the ability of the mortgagors to service their debts. I now turn to the market anticipations of housing prices: the methods used and why they were so erroneous.

Gerardi et al. (Brookings Papers) explore whether market participants could have or should have anticipated the large increase in foreclosures that occurred in 2007. They decompose the change in foreclosures into two components: the sensitivity of foreclosures to a change in housing prices times the change in housing prices. The authors conclude that investment analysts had a good sense of the sensitivity of foreclosures to a change in housing prices, but missed drastically the expected change in housing prices. The authors do not analyze whether housing was overvalued in 2005–2006 or whether the housing price change was to some extent predictable.

The authors looked at the records of market participants from 2004 to 2006 to understand why the investment community did not anticipate the subprime mortgage crisis. Several themes emerge. The first is that the subprime market was viewed as a great success story in 2005. Second, mortgages were viewed as lower risk because of their more stable prepayment behavior. Third, analysts used sophisticated tools but the sample space did not contain episodes of falling prices. Fourth, pessimistic feelings and predictions were subjective and not based upon quantitative analysis. Hence, they were disregarded.

Analysts were remarkably optimistic about Housing Price Appreciation (HPA). Those who looked at past data on housing prices, such as the four-quarter appreciation, could construct the histogram below. In the aggregate, housing prices never declined from year to year during the period 1980q1–2007q4. The mean appreciation was 5.4% pa with a standard deviation of 2.94% pa. The optimism could be understood if one asks: on the basis of this sample of 111 observations, what is the probability that housing prices will decline? Given the mean and standard deviation, there was only a 3% chance that prices would fall.

The best estimates of the analysts were that the rates of housing price appreciation CAPGAIN or HPA in 2005–2006 of 10–11% per annum would be unlikely to be repeated but that it would revert to its longer term average. A Citi report in December 2005 stated that "...the risk of a national decline in home prices appears remote. The annual HPA has never been negative in the United States going back at least to 1992." Therefore no mortgage crisis was anticipated. There was no economic theory or analysis in this approach. It was simply a VaR value at risk implication from a sample based upon relatively recent data. More fundamentally, no consideration was given to the economic determinants of the probability distribution of capital gains or housing price appreciation (Fig. 2.1).

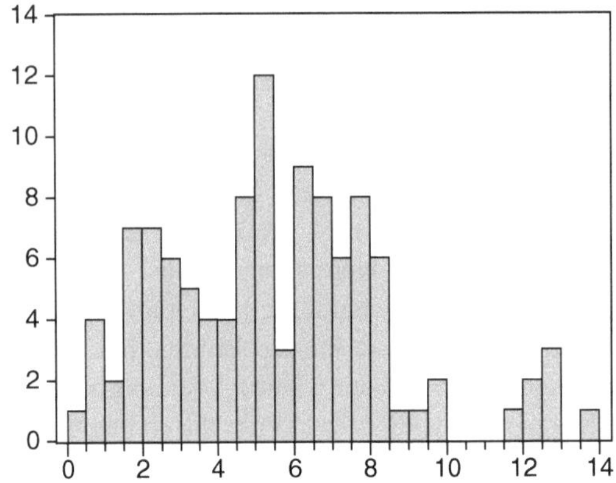

Series: CAPGAIN	
Sample 1980Q1 2007Q4	
Observations 111	
Mean	5.436757
Median	5.220000
Maximum	13.50000
Minimum	0.270000
Std. Dev.	2.948092
Skewness	0.562681
Kurtosis	3.187472
Jarque-Bera	6.019826
Probability	0.049296

Fig. 2.1 Histogram and statistics of CAPGAINS = Housing Price Appreciation HPA, the change from previous four-quarter appreciation of US housing prices, percent/year, on horizontal axis. Frequency is on the vertical axis (Source of data: Office of Federal Housing Price Oversight)

2.1.4 The Disregarded Warnings

Greenspan, Bernanke and the IMF were insouciant, but there were Cassandras who warned of the housing price bubble and likelihood of a collapse. Shiller (2007) looked at a broad array of evidence concerning the recent boom in home prices, and concluded that it does not appear possible to explain the boom in terms of fundamentals such as rent and construction costs. Instead he proposed a psychological theory or social epidemic. This "explanation" is not convincing theoretically, and was not able to overcome the Jackson Hole Consensus. One can do much better than invoke vague phrases such as "epidemic", "contagion", "irrationality".

From 1998 to 2005 rising home prices produced above average capital gains, which increased owner equity. This induced a supply of mortgages, and the totality of household financial obligations as a percent of disposable personal income rose. Figure 2.2 graphs the ratio of housing prices/disposable income P/Y = PRICEINC and the debt service DEBTSERVICE, which is interest payments/disposable income. In Fig. 2.2, both variables are normalized, with a mean of zero and standard deviation of one.

The rises in housing prices and owner equity induced a demand for mortgages by banks and funds. In about 45–55% of the cases, the purpose of the subprime mortgage taken out in 2006 was to extract cash by refinancing an existing mortgage loan into a larger mortgage loan. The quality of loans declined. The share of loans with full documentation substantially decreased from 69% in 2001 to 45% in 2006 (Demyanyk and Van Hemert 2007). The ratio of debt/income rose drastically. The only way to service or refinance the debt was for the capital gain to exceed the mean interest rate. This is an unsustainable situation since it implies that there is a "free lunch" or that the present value of the asset diverges to infinity.

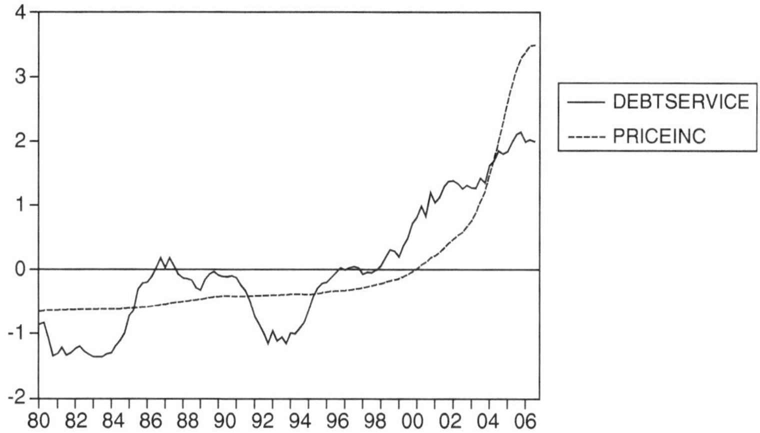

Fig. 2.2 PRICEINC = Ratio of housing prices/disposable income. DEBTSERVICE = Debt service/disposable income. Both variables are normalized (FRED data set of the Federal Reserve Bank of St. Louis, Office of Federal Housing Enterprise Oversight)

The fatal error was to ignore the fact that the quality of mortgages declined and that it was ever less likely that the mortgagors could service their debt from current income. Sooner or later the defaults would affect housing prices and turn capital gains into capital losses. The market gave little to no consideration of what would happen if the probability distribution/histogram would change. Both the supporters and the critics of the Jackson Hole Consensus agree that asset price bubbles are a source of danger to the real economy if the financial structure is fragile and not properly capitalized. The danger from "overvaluation" of housing prices is that the debt used to finance the purchase is excessive, which would lead to defaults and foreclosures.

It is seen in Fig. 2.2 that the ratio PRICEINC = P(t)/Y(t) and the DEBTSERVICE ratio were stable, almost constant from 1980 almost to 2000. Then there was a housing bubble, the price/income shot up from 2000 to 2006. As a result of the rise in homeowner's equity the debt ratio rose – to finance consumption. The debt service ratio rose to two standard deviations above the longer term mean. The great deviation of the price/income ratio from its long term mean would suggest that there was a housing price "bubble" and that housing prices were greatly overvalued. A housing crisis would be predicted, when the ratio P(t)/Y(t) would return to the long term mean, which is the zero line. Households would then default on their mortgages and leverage would transmit the shock to the financial sector. The market – as well as the Fed – discounted that apprehension. There was no theory that could identify an asset price bubble and its subsequent effect upon the economy. The Jackson Hole Consensus ignored the microeconomy.

There were financial firms who may have had qualms about the sustainability of the housing price appreciation, but they assumed that they would be able to anticipate the onset of a crisis in time to retrench. Charles Prince's remark is emblematic: "When the music stops, in terms of liquidity, things will be complicated. But as long as the music is playing, you've got to get up and dance". They certainly were mistaken, because they ignored systemic risk that the negative shock could be pervasive, and liquidity and capital would disappear in the wake of a mass exodus from the markets for derivatives.

There were a few hedge firms such as Scion Capital (SC) that anticipated the crash and took appropriate actions. Michael Burry (2010) of SC realized in 2005 that the bubble would burst and acted upon that view. He purchased credit default swaps (CDS) on billions of dollars worth both of subprime mortgage backed securities and bonds of many financial corporations that would be devastated when the real estate bubble burst. Then as the value of the bonds fell, the value of CDS would rise. The investors in his hedge fund still "wanted to dance" and profit from the rising house prices. Despite pressure from the investors, Burry liquidated the CDS at a substantial profit. But since he was operating in face of strong opposition from both his investors and from the Wall Street community, he shut down SC in 2008. Greenspan responded negatively to Burry's predictions and suggested that Burry was just lucky. Lowenstein (2010, Chap. 7) describes the divergent opinions in the market where the pessimists were in the minority.

2.1.5 The Controversy Over Regulation and Deregulation

In the years prior to the crisis there was a controversy concerning the regulation of Over the Counter (OTC) markets for derivatives in particular. The warnings and issues raised were disregarded and strongly opposed by the Fed and Treasury. The warnings were public information. After the crisis, the disregarded policies were embodied in the Dodd-Frank bill. The errors of Greenspan, Rubin were of those of commission not omission.

I draw upon the FCIC report to explain the controversy. Derivative markets are organized as exchanges and as OTC markets. Exchanges are regulated by Federal law and play an important role in the price discovery process. They reveal as public information market anticipations of commodity prices and of interest rates underlying futures and derivatives. See Stein (1986) for the economics of futures markets. OTC derivatives are traded by large financial institutions – traditionally bank holding companies, investment banks – which act as derivatives dealers, buying and selling contracts with the customers. Unlike futures and options exchanges, the OTC market is neither centralized nor regulated. The price discovery process is limited. The OTC market is opaque.

In May 1998 Commodities Futures Trading Commission (CFTC) chairman Brooksly Born said that the agency would reexamine the way it regulated the OTC derivatives market. The US General Accounting Office (GAO) issued a report on financial derivatives that found systemic risks and dangers in the concentration of OTC derivative activity among 15 major dealers. The GAO concluded that "the sudden failure or withdrawal of any one of these dealers could cause liquidity problems in the markets and could also pose risks to the others, including federally insured banks and the system as a whole".

The Treasury, Fed and SEC – Greenspan, Rubin, Levitt and Summers opposed regulation. Greenspan said "…regulation of derivatives transactions that are privately negotiated by professionals is unnecessary". He continued to advocate deregulation of the OTC market and the exchange traded market. Greenspan believed that: "By far the most significant event in finance during the past decade has been the extraordinary development and expansion of financial derivatives."

In 1999 the President's Working Group on Financial Markets, a committee of the heads of the Treasury, the Federal Reserve, SEC and CFTC charged with tracking the financial system, chaired by Summers, essentially adopted Greenspan's view. The group issued a report urging Congress to deregulate OTC derivatives broadly and reduce CFTC regulation of exchange traded derivatives as well. In December 2000, in response, Congress passed and President Clinton signed the Commodities Futures Modernization Act (CFMA) of 2000. This act in essence deregulated the OTC derivatives market and eliminated oversight, regulation by both the CFTC and SEC.

The Act let derivatives traders, including large banks and investment banks, increase their leverage. A key OTC derivative in the financial crisis was the Credit Default Swap (CDS). This is a type of "insurance" against capital losses on assets. I discuss the role of CDS in Chap. 6. After the CFMA was passed, the N.Y. State Insurance Department determined that "naked" CDS did not count as insurance and therefore was not subject to regulation.

When an insurance company sells a policy, insurance regulators require that the company set aside reserves in case of loss. In the housing boom, CDS were sold by firms that failed to put up any reserves, or collateral to hedge their exposure. In the run up to the crisis of 2007–2008, AIG – the largest US insurance company – accumulated one-half a trillion dollar position in credit risk through the OTC market without being required to post even one dollar of initial capital or making any other provision for loss. AIG was not alone.

The value of the underlying assets for CDS outstanding worldwide grew from $6.4 trillion at the end of 2004 to $58.2 trillion at the end of 2007. A significant portion was apparently speculation in "naked" CDS. In addition, much of the risk of CDS and other derivatives were concentrated in a few of the very large banks and others such as AIG Financial Products.

Among those who warned of systemic problems was Sheila Blair, Chairman of the FDIC. She focused upon the ultimate source of the problem – the mortgages – and the banking sector. She said: "Subprime lending was started and the lion's share occurred in the nonbank sector. But it clearly created pressures on the banks…I think nipping this thing in the bud in 2000 and 2001 with some strong consumer rules applying across the board that just simply said you've got to document a customer's income to make sure they can repay the loan…just simple rules like that". At the Fed, it was well known that Greenspan was not interested in increased regulation.

The danger of systemic risk, resulting from the financial structure of derivatives, was public information that was disregarded by the markets, the Fed and Treasury and the regulators. I discuss these issues in Chap. 3 on the Quants.

Ex-post the Dodd-Frank (D-F) bill tried to rectify the earlier disregard of the warnings. The main provisions of the D-F bill are as follows. First, it aims to create transparency and accountability for derivatives. Second, it aims to close "Regulatory Gaps". It provides the SEC and CFTC with authority to regulate over-the-counter derivatives so that excessive risk-taking can no longer escape regulatory oversight. Third, it requires central clearing and exchange trading for derivatives that can be cleared and provides a role for both regulators and clearing houses to determine which contracts should be cleared. Fourth, it concerns market transparency. It requires data collection and publication through clearing houses or swap repositories to improve market transparency and provide regulators important tools for monitoring and responding to risks. Fifth, it concerns financial safeguards. It adds safeguards to system by ensuring that dealers and major swap participants have adequate financial resources to meet responsibilities, and provides regulators the authority to impose capital and margin requirements on swap dealers. It aims to ends the possibility that taxpayers will be asked to write a check to bail out financial firms that threaten the economy, by creating a safe way to liquidate failed financial firms and imposing new capital and leverage requirements that make it undesirable to get too big. It updates the Fed's authority to allow system-wide support but no longer props up individual firms. It establishes rigorous standards and supervision to protect the economy and American consumers, investors and businesses.

Finally, it seeks an advance warning system. It creates the Financial Stability Oversight Council to identify and address systemic risks posed by large, complex companies, products, and activities before they threaten the stability of the economy. The Council is chaired by the Treasury Secretary and includes the Federal Reserve Board, SEC, CFTC, OCC, FDIC, the new Consumer Financial Protection Bureau, and an independent appointee with insurance expertise. Nonvoting members include state banking, insurance, and securities regulators. The D-F bill attempts to reverse the policies of Geenspan, the Fed and Rubin.

2.1.6 The Failures of International Monetary Fund Surveillance

The Independent Evaluation Office (IEO 2011) of the IMF found that the IMF provided few clear warnings about the risks and vulnerabilities associated with the crisis before its outbreak. I basically quote directly from this important report. The IMF economists tended to hold in the highest regard macro models of the academic economics profession. These models proved inadequate for analyzing financial linkages. Specifically, they introduced money and asset markets in the most rudimentary way, completely unrelated to economic reality. The failure of the IMF surveillance, discussed in this section, is in fact prima facie evidence of the failure of the dominant macroeconomic models.

The Fund missed key elements that underlay the developing crisis. In the US it did not discuss, until the crisis already erupted, the deteriorating lending standards for mortgage financing nor did the Fund adequately assess the risks and impact of a major housing price correction on highly leveraged financial institutions. As late as April 2006, shortly before US housing prices peaked, the Fund Surveillance reports explained away the rising share of non-traditional mortgages in the US by writing "Default rates on residential loans have been low historically. Together with the securitization of the mortgage market, this suggests that the impact of a slowing housing market on the financial sector is likely to be limited."

The Fund's reading of the financial sector was remote from reality. In 2007 the Fund concluded that the financial sector was sound: "Core commercial and investment banks are in a sound financial position, and systemic risks appear to be low. Profitability and capital adequacy of the banking system are high by international standards....market measures of default risk have remained benign".

A remarkable failure of the Fund concerns Iceland before its major financial crisis. In spite of a banking sector that had grown from about 100% of GDP in 2003 to almost 1,000% in 2007, this was not noted as a vulnerability that needed to be addressed urgently. The staff reports about Iceland were sanguine. The report stated "Iceland's medium-term prospects remain enviable." In fact, "the banking sector appears well-placed to withstand significant credit and market shocks".

The IEO explains why the IMF failed to give clear warnings. There was *groupthink*, a tendency among homogeneous cohesive groups to consider issues only within a certain paradigm and not challenge its basic premises. They were in awe of the dominant macroeconomic models mentioned above. The basic premises were that market discipline and self-regulation would be sufficient to stave off serious problems in financial markets. Crises were unlikely to happen in advanced economies where sophisticated financial markets could thrive with minimal regulation. The Fund accepted the same inappropriate macroeconomic model that dominates the academic profession.

Added to groupthink was the *confirmation bias* which is the tendency of people to only notice information consistent with their own expectations and ignore information that is inconsistent with them.

2.2 Conclusions

The crucial ultimate variable is the household debt, the mortgage debt. The rest of the financial system rested upon the ability of the mortgagors to service their debts. *Systemic risk* relates the effects of the failure of the mortgagors to service their debts upon the financial structure based upon the income from mortgages. The leverage of the financial system transmitted the housing market shock into a collapse of the financial system.

A bubble is in effect a large positive unsustainable excess debt. Detection of a bubble corresponds to the detection of an "excess debt". The latter is equal to the difference between the actual and the optimal debt. The fundamentals are reflected in the optimal debt. The aim of this book is to derive an optimal leverage, debt or capital requirement. The technique of analysis is Stochastic Optimal Control (SOC). This analysis balances the expected growth against risk. Thus an excess debt corresponds to a situation where asset prices deviate from the derived fundamentals. There is no use of an arbitrary jump process where a crash occurs. Instead the SOC analysis implies that the probability of a crisis is directly related to the measured excess debt.

The housing price bubble, its subsequent collapse and the financial crisis were not predicted either by the market, the Fed, the IMF or the regulators in the years leading to the current crisis. Warnings, based upon publicly available information, were not only disregarded but the Fed and Treasury rejected the warnings. They successfully advocated deregulation of OTC markets. As a result, transparency of prices was reduced, risk was concentrated in a few major financial institutions and high leverage was induced. These were basic ingredients for the subsequent crisis.

The Fed, the IMF and Treasury lacked adequate tools which might have indicated that asset values were vastly out of line with fundamentals. The Fed and the Fund were not searching for such tools because they did not believe that they could or should look for misaligned asset values or excess debt, despite warnings from Shiller, some people in the financial industry, the GAO, state bank regulators and FDIC. The Fed was blind-sided by the Jackson Hole Consensus which gave them great comfort in adopting a hands off position. So it was not just a lack of appropriate tools which undid the Fed; it was a complete lack of appreciation of what its role should be to head off an economic catastrophe. Peter Clark wrote that there are two separate but related questions: Are identification and containment of a financial bubble legitimate activities of the Fed, and if they are, what are the best tools to carry out this analysis? As the Fed answered "No" to the first question, it saw no need to address the second question.

References

Bernanke, Ben. 2005. *Testimony to congress joint economic committee*, October.

Burry, Michael J. 2010. I saw the crisis coming. Why didn't the fed?. *New York Times* April 4, 2010, Op/Ed page.

Clark, Peter. 2009. Comment on tale of two debt crises. *Economics, The Open-Assessment E-Journal*, http://www.economics-ejournal.org/economics/discussionpapers/2009–44.

Congressional Oversight Panel. 2009. *Special report on regulatory reform*, Washington DC, January.

Demyanyk, Yulia, and Otto Van Hemert. 2007. Understanding the subprime mortgage crisis. SSRN: <http://ssrn.com/abstract=1020396>

Financial Crisis Inquiry Report, FCIC Final Report of National Commission. 2011.

Fleming, Wendell H., and Jerome L. Stein. 2004. Stochastic optimal control, international finance and debt. *Journal of Banking and Finance* 28: 979–996.

FRED, Federal Reserve Bank St. Louis, Economic Data Set.

Gerardi, K., A. Lehnert, S. Sheelund, and P. Willen. 2008. Making sense of the subprime crisis. *Brookings Papers*, Fall.

Greenspan, Alan. 2004a. *Understanding household debt obligations*. Washington, DC: Remarks at Credit Union National Association.

Greenspan, Alan. 2004b. *The mortgage market and consumer debt*. Washington, DC: Remarks at America's Community Bankers Annual Convention.

Greenspan, Alan. 2008. Testimony on sources of financial crisis. *House committee on government oversight and reform*.

Greenspan, Alan. 2010. The crisis. *Brookings Papers*.

International Monetary Fund. 2007. World Economic and Financial Survey, Global Financial Stability Report.

Issing, Otmar. 2009. *Asset Prices and Monetary Policy*. Cato Journal, 29.

Koopmans, T., and A. Bausch. 1959. Selected topics in economics involving mathematical reasoning. *SIAM Review* 1(2): 79–148.

Lowenstein, Roger. 2010. *The end of Wall Street*. New York: Penguin Press.

OFHEO. 2010. House price indexes, Washington, DC.

Shiller, Robert. 2007. Understanding recent trends in house prices and home ownership. Cowles foundation discussion paper #1630.

Stein, Jerome L. 1986. *Economics of futures markets*. New York: Blackwell.

Chapter 3
Failure of the Quants

Abstract The Quants made several interrelated errors. They based the expected house price upon the current price and drift, and assumed that the distribution of capital gains is unchanging. The Quants ignored publicly available information in forming expectations. The ultimate determinant of the values of CDOs was the ability of the mortgagors to service their debts. It was public information that the mortgages were of most dubious value, no due diligence was performed in the ratings and the debt/income of the mortgagors was rising. Hence it was improbable that the distribution of house prices would remain constant. Systemic risk was ignored by the Quants who just focused upon current prices. Ratings of the tranches were not based upon the quality of the underlying mortgages. They were all in the same bundle. The rating depended upon who got paid first in the stack of loans. The key question was how to rate and price the tranches. The issue concerned the correlation of the tranches. If a pool of loans started experiencing difficulties, and a certain percent of them defaulted, what would be the impact upon each tranche? The "apples in the basket model" made one prediction. Another very different one is "the slice of bread in the loaf" model.

This chapter considers the role of the "Quants" in the crisis. The financial system is an intermediary between savers and investors. Mathematics is essential to price securities traded by savers, intermediaries and investors, and to adjust risk to the preferences of savers and investors. Assets are priced according to the principle that the price should be equal to the expectation of the present value of the future income. The future income stream and time profile of future interest rates are crucial variables, but they are stochastic and unknown when the pricing decision is made. Similarly the pricing of longer-term bonds depends upon the expectations of future short term interest rates. This pricing is based upon the term structure hypothesis that the long rate is a geometric average of future short rates.

J.L. Stein, *Stochastic Optimal Control and the U.S. Financial Debt Crisis*, DOI 10.1007/978-1-4614-3079-7_3, © Springer Science+Business Media New York 2012

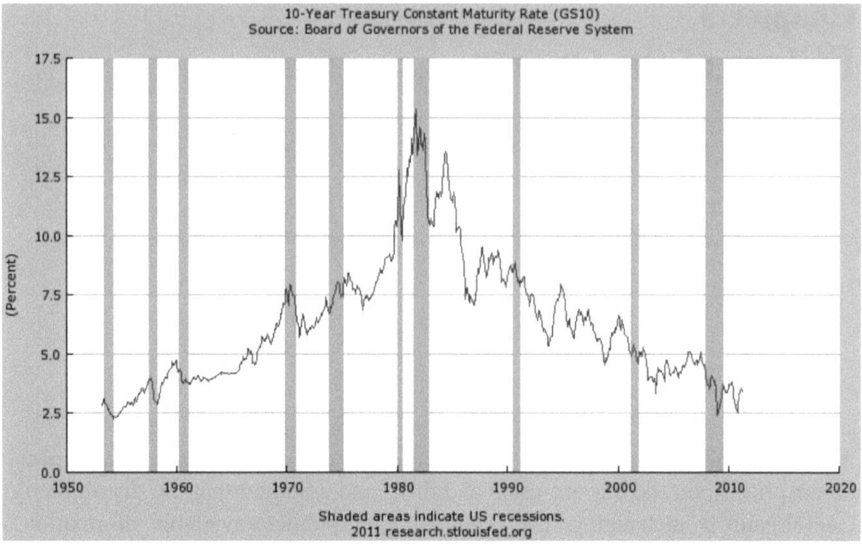

Fig. 3.1 Ten-year treasury constant maturity rate

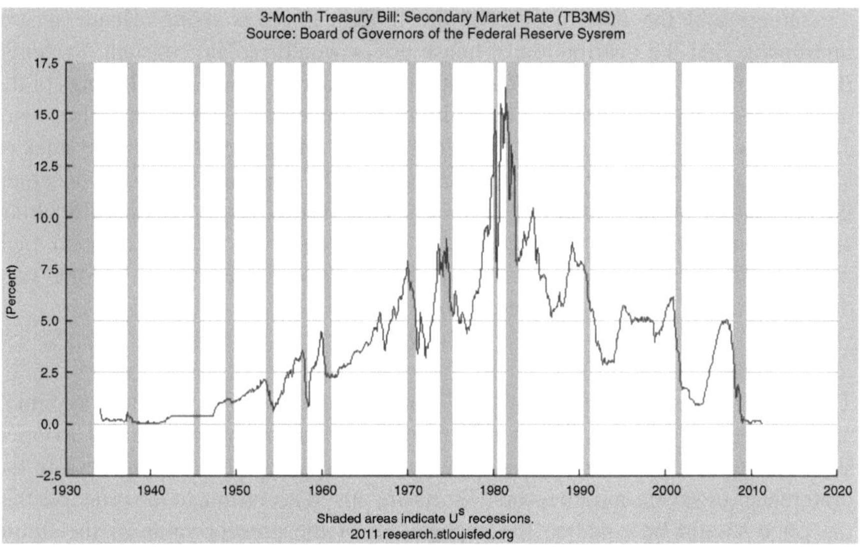

Fig. 3.2 Three-month Treasury bill rate

When the Federal Reserve changed its operating policy in 1979, interest rates became more responsive to fluctuations in aggregate demand and supply shocks. Figure 3.1 graphs the 10-year Treasury Constant Maturity Rate and Fig. 3.2 graphs the 3-month Treasury bill rate. The variability of the Treasury bond rate carried over to the mortgage market is graphed in Fig. 3.3.

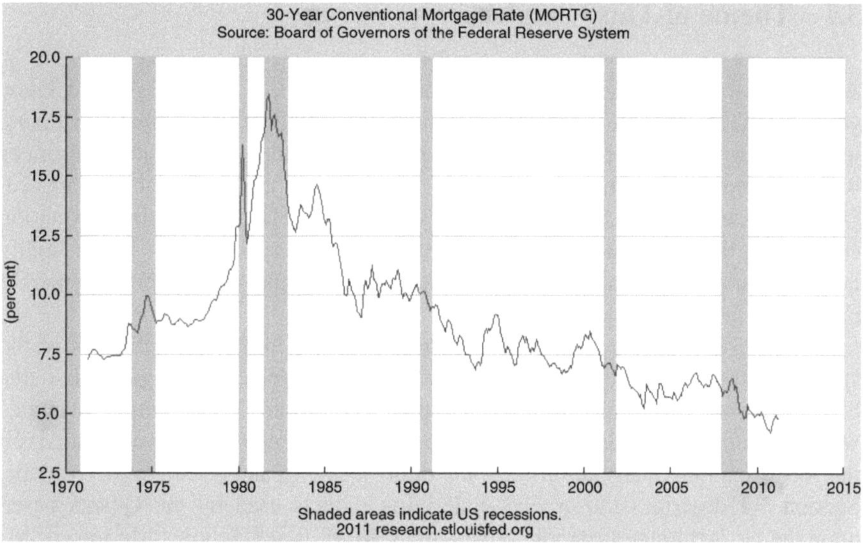

Fig. 3.3 Thirty-year conventional mortgage rate

As a result of the considerably increased volatility of both short and long term interest rates there was an increased the demand for risk management. One took a great speculative risk if a bond position was not hedged.

In the volatile economic climate of the 1970s, trading in financial futures was introduced. Between 1970 and 1984 futures trading volume increased tenfold and by the end of 1984 financial futures accounted for 50% of total futures trading. The Federal Reserve Board (FRB), Securities Exchange Commission (SEC) and Commodities Futures Trading Commission (CFTC) were directed by an act of Congress in 1982 to carry out a joint study of the economic purposes of futures and options markets, to consider their effects upon capital formation, and to evaluate the adequacy of regulation. I was commissioned to write on the first topic which later resulted in my book *The Economics of Futures Markets* (1986). My work seemed to satisfy the FRB, SEC and CFTC that these markets served useful economic functions and were not "gambling casinos".

The interest rate volatility (Figs. 3.1, 3.2, 3.3) above generated a demand by Wall Street for Quants. The Quants are a group of physicists, mathematicians and computer science experts who practice "financial engineering". They develop and apply models to devise derivatives that would permit the risks to be hedged to an optimal extent, making the new securities attractive to both savers and investors. The channeling saving to investment is important for the growth process.

3.1 Theme of This Chapter

According to the Efficient Market Hypothesis (EMH), it is impossible to know when a bubble is occurring, since current prices reflect all publicly available information. Only in retrospect, when the bubble bursts is it clear that the prices were inflated. I show that the EMH did not describe accurately what happened. Quants missed systemic risk because they ignored publicly available information concerning the mortgage market. That market was very important in the highly leveraged and interrelated financial markets. The collapse of the mortgage market bubble led to the financial collapse.

Section 3.2 is a discussion of leveraging, and leads into a vivid example of the strategy of the Atlas Fund run by Quants, which was given AAA ratings. At first, this hedge fund was extremely profitable and then it collapsed. This example illustrates why the Quants failed. Section 3.3 describes the structure of the derivatives market, the securitization, interrelationships, the rating agencies and pricing of derivatives. Section 3.4 describes the crucial underlying models used by the Quants based upon the No Arbitrage Principle NAP – the CAPM, Black-Scholes-Merton (BSM) and EMH – in devising their instruments and strategy. Section 3.5, discusses the models, based upon the No Arbitrage Principle NAP, in detail. Section 3.6 concerns methods to determine when the drift of the capital gains has changed. The conclusion is Sect. 3.7. It summarizes the errors made by the Quants in the pricing of the tranches and the Credit Default Swaps. The next few chapters explain why Stochastic Optimal Control (SOC) uses information much better and can provide Early Warning Signals (EWS) of a crash.

3.2 Leveraging

It is now widely believed that "excessive" leveraging, and an "excessive" debt ratio, at key financial institutions helped convert the initial subprime turmoil in 2007 into a full blown financial crisis of 2008. Leverage is the ratio of assets /net worth $A(t)/X(t)$, or equivalently the ratio of debt $L(t)$ to net worth $X(t)$ denoted $f(t) = L(t)/X(t)$. Although leverage is a valuable financial tool, "excessive" leverage poses a significant risk to the financial system. For an institution that is highly leveraged, changes in asset values greatly magnify changes in net worth. To maintain the same debt ratio when asset values fall either the institution must raise more capital or it must liquidate assets.

The relations are seen through Equations (i)–(iv). In (i) net worth $X(t)$ is equal to the value of assets $A(t)$ less debt $L(t)$. Equation (ii) is just a way of expressing the debt ratio. Equation (iii) relates the debt ratio $f(t) = L(t)/X(t)$ to the leverage ratio $A(t)/X(t)$ of assets/net worth. Equation (iv) states that the percent change in net worth $dX(t)/X(t)$ is equal to the leverage $(1 + f(t))$ times $dA(t)/A(t)$ the percent change in the value of assets.

Table 3.1 Leverage of institutions

Broker-dealers and hedge funds	27
Government sponsored enterprises	17
Commercial banks	9.8
Savings banks	6.9
Average, weighted	12

(i) $X(t) = A(t) - L(t)$.
(ii) $L(t)/X(t) = f(t) = 1/[(A(t)/L(t) - 1]$.
(iii) $A(t)/X(t) = 1 + f(t)$.
(iv) $dX(t)/X(t) = (1 + f(t))\ dA(t)/A(t)$.

The Congressional Oversight Panel (2009) (COP) reported that, on the basis of estimates just prior to the crisis, investment banks, securities firms, hedge funds, depository institutions, and the government sponsored mortgage enterprises – primarily Fanny Mae and Freddie Mac – held assets worth \$23 trillion on a base of \$1.9 trillion in net worth, yielding an overall average leverage of $A/X = 12$. The leverage ratio varied widely as seen in (Table 3.1).

Consider the average leverage, where $A(t) = \$23$ trillion, $X(t) = \$1.9$ trillion, $L(t) = \$21.1$ trillion, making the debt ratio $f = 11.1$. From Equation (iv), a 3% decline in asset values would reduce net worth by $dX(t)/X(t) = (1 + 11.1)(0.03) = 36\%$. The loss of net worth is equal to (0.36) (\$1.9 trillion) = \$0.69 trillion. To maintain the same ratio $f = 11$, the institutions must either raise capital to offset the decline in asset values $dX = dA < 0$, or must sell off assets to reduce debt by the same proportion $dL(t)/L(t) = dX(t)/X(t)$, derived from Equation (ii). A 3% decline in asset value would require the sale of (0.03) (21.1 trillion) = \$630 billion in assets to repay the debt.

Both actions have adverse consequences for the economy. Firms in the financial sector, the financial intermediaries, are interrelated as debtors-creditors. Banks lend short term to hedge funds who invest in longer term assets and who may also buy credit default swaps. Firms that lost \$690 billion in net worth would have difficulty in raising capital to restore their net worth, without drastic declines in share prices. Similarly, the attempt by one group to sell \$630 billion in assets to repay loans will have serious repercussions in the financial markets. The prices of these assets will fall, and the leverage story repeats for other groups. Institutions who hold these assets will find that the value of their portfolios have declined, reducing their net worth. In some cases, there are triggers. When the net worth of a Fund falls below a certain amount ("breaks the buck") the fund must dissolve and sell its assets, which may include AAA assets. In turn the sale of AAA assets affects other institutions. Conservative investors may have thought they were holding very safe assets, but to their dismay they suffer capital losses as AAA assets are liquidated at "fire sale" prices. In a highly interrelated system, "high leverage" can be very dangerous. What seems like a small shock in one market can affect via leverage the whole financial sector. The Fed and the IMF seemed oblivious to this systemic risk phenomenon because of the history of two previous bubbles.

In the S&L and agricultural crises of the 1980s, discussed in Chap. 7, there was not a strong linkage between the specific sector and a highly leveraged interrelated financial sector based upon CDO and CDS. Therefore the collapse of these earlier bubbles only had localized effects.

3.2.1 The Incredible Leverage of Atlas Capital Funding

The story of the Atlas Capital Fund is an excellent example of leveraging discussed above. My discussion is based upon a paper given by Jichuan Yang, one of the principals of Atlas, given at an Applied Mathematics Colloquium at Brown University in September 2009 and the paper by Ren Cheng (former Chief Investment Officer at Fidelity) at the same Colloquium. A group of talented financial engineers: mathematicians, physicists specializing in mathematical finance, decided to establish a Fund in 2003 with $12 *billion* of assets, and $10 *million* of capital, – a leverage of 1,200. This Fund was called the Atlas Capital Fund, due to its huge size. The fund portfolio would contain thousands of individual bonds, loans and other financial securities that had longer term maturities, such as 8 years. The liabilities were commercial paper and mid-term notes with maturities ranging from 30 days to 5 years. Atlas would borrow short term and lend longer term to the Hedge Funds. The Fund was not set up to hedge risk but to seek maximum return. The Fund did not fear taking risk. Atlas would make its profits from the difference between the lending rate charged to the hedge funds and the cost of short term borrowing. The latter could be reduced to a minimum if Atlas received a AAA rating. This was a remarkable goal because most global banks are rated no higher than AA.

Since the portfolio had a much longer maturity than the loans, a major risk to Atlas would be the variable short term borrowing rate. Figure 3.2 graphs the volatility of the short term interest rate. When the 30-day loan matured, Atlas would roll over the 30-day loan at the current rate. If there were difficulties in rolling over, Atlas would have to find banks, called "liquidity providers", to give Atlas "emergency" loans to pay off the 30-day debt.

The "financial engineers" built a model to evaluate the risk, which they used to convince the rating agencies to give them an AAA rating, which lowers the cost of borrowing. The model simulated the movement of the $12 billion of individual assets as well as their correlated behavior. The mismatch of the timing of cash flows of assets and liabilities, the price movements, the rating changes, the defaults and recovery had to be "accurately" modeled, calculated and simulated. For each potential future price movement, the model calculated the gain, loss and return. After tens of thousands of such simulations, the financial engineers arrived at an estimate of the expected loss and expected return by certain types of averaging the individual outcomes. These simulations convinced the rating agencies to give Atlas an AAA rating and hence a low cost of borrowing.

At first, Atlas was extremely profitable. Stockholders received 100% of their money back in the first year of operations due to the leverage of 1,200, equal to $12 billion of assets/$10 million of capital. The Fed was most accommodating with its low interest policy. Moreover, Chairman Alan Greenspan was the champion of financial innovation and was fighting off regulatory reform on all fronts, as discussed in Chap. 2. About 3 years after Atlas started operations, the US financial industry went into one of its worst crises. The cascading effects of leverage discussed above then occurred. Atlas was blamed as being one of the main culprits causing the crisis. Jichuan Yang, a principal of Atlas, wrote in 2009: "Today, if someone tells me that all these things can be simulated by an elegant mathematical model with any realistic accuracy, I would be tempted to say that he's probably an overconfident idiot".

3.3 Structure of Derivatives Market, Rating Agencies and Pricing of Derivatives

This part draws upon the comprehensive FCIC report, and the books by Derman and by Patterson. The financial market consisted of several stages. At one end were the mortgagors, the households who borrowed against negligible collateral or ability to service the loans from income, the NINJAs (no income no jobs, no assets). Their loans were packaged or securitized into bundles by financial intermediaries who could not perform due diligence because they had no idea of the quality of the loans. In turn, these packages were sold to institutional investors, who relied upon the rating agencies to evaluate risk. This method was used by FNMA and then followed by the private sector.

Vast quantities of similar but not identical securities were pooled into bundles and sold to large investors, mutual funds, pension funds, insurance companies and hedge funds who were seeking high return and low risk investments. There was no clear way to evaluate the risk of a package, because it consisted of many mortgages of dubious quality. However, it was believed that by pooling the securities the risk would be diversified. There would be losses on the poor quality ones but the returns on the high quality mortgages would be uncorrelated with the poorly performing group. That is, the bundle would be like *apples in a basket*. There will be some rotten ones, but there would also be good ones. The process of asset acquisition, pooling and standardization would be attractive to investors. There was a belief in geographic diversification and that real estate markets are local ones/independent. This belief reflected the fact that nationwide real estate price declines had not been experienced since the Great Depression. The banks, the original owners of the mortgages, sold them to the investment houses that securitized the packages, and could make more loans. The supply of mortgages would then be increased, just as the demand was increased by the risk diversification packages. Greenspan believed that: "By far the most significant event in finance during the past decade has been the extraordinary development and expansion of financial derivatives."

The Quants devised a method to sell the derivatives of dubious value as AAA securities. A prioritization method was used. Tranches of the pool were sold, like bonds, preferred stock and common stock *in the same bundle*. Losses or defaults would first affect the lowest tranche (Equity) then when that tranche's assets were exhausted, the losses would affect the medium (Mezzanine) tranche. The top, senior tranche (AAA) would only lose if the losses exceeded the assets in all the tranches.

Ratings were not based upon the quality of the underlying mortgages all of which were in the same bundle. The rating depended upon who took losses first in the stack of loans. The rating agencies were essential to the smooth functioning of the mortgage backed securities market. Banks needed their ratings to determine the amount of capital to hold, repo markets needed the ratings to determine loan terms, and some investors could only buy securities with a AAA rating. Credit ratings also determined whether investors could buy certain investments. The SEC restricts money market funds to purchasing securities that have received credit ratings in one of the two highest short-term rating categories. The Secondary Mortgage Market Enhancement Act of 1984 permitted federal and state chartered financial institutions to invest in mortgage related securities if the securities had high credit ratings from at least one rating agency. Many investors, such as pension funds and university endowments relied on credit rating agencies because they had neither access to the same data as the rating agencies nor analytical ability to assess the securities that they were purchasing.

The logic of trenching was the *apples in a basket model*. The *loaf of bread* model, discussed in my criticism below, was not considered. Say that there were two tranches A(AAA, top) and B (Equity, bottom). Let $1 > p > 0$ be the probability of default of the entire package. The losses first affect tranche B. Tranche A can only suffer losses if tranche B is wiped out.

The probability Pr(B) of default of tranche B is $1 > p > 0$. The probability that that both tranches suffer losses is Pr(AB). In the *Apples in a Basket* case the defaults in the two tranches are *independent samples* of apples from a population (basket) where the probability of default is p. A first apple (Equity or Mezzanine tranche) is taken from the basket and it turns out to be rotten. This will occur with probability p. A second apple (Senior tranche) is taken out of the basket. In the "basket of apples" case, the $\Pr(A|B) = \Pr(B) = p$. In this case of independence, the probability that both apples are rotten corresponds to the case where the senior tranche would be exhausted $\Pr(AB) = \Pr(A)\Pr(B) = p^2 < p$. Risk of the senior tranche is less than that of the entire package. The senior tranche appeared to be a silk purse made from a sow's ear.

Since the mid-1990s Moody's has rated tranches of mortgage backed securities using several models. Although Moody's did not sample or review individual loans, the company used loan/value ratios, borrower credit scores and loan terms. The model simulated the performance in 1,250 scenarios including variations in interest rates and state unemployment rates as well as in home price changes. Thereby ratings were given to the tranches.

On average across the scenarios, home prices trended upwards at approximately 4% per annum. The model put little weight on the possibility that prices would fall sharply nationwide. Even as house prices rose to unprecedented levels, Moody's never adjusted the scenarios to put greater weight on the possibility of a decline. This choice did not consider that there was a national housing bubble and did not sufficiently account for the deterioration of the quality of loans securitized. In October 2007, Moody's downgraded all of the mortgage backed securities that it had rated AAA in 2006, and downgraded 73% to junk. There was indeed a house price bubble. The obvious question is why was the rating system a failure?

3.3.1 Pricing CDOs

Ratings of the tranches were not based upon the quality of the underlying mortgages. They were all in the same bundle. The rating depended upon who got paid first in the stack of loans. The key question was how to rate and price the tranches. The issue concerned the correlation of the tranches. If a pool of loans started experiencing difficulties, and a certain percent of them defaulted, what would be the impact upon each tranche? The "apples in the basket model" made one prediction. Another very different one is "the slice of bread in the loaf" model.

The *slices of bread* model assumes systemic risk, in which the value/returns of assets are very highly or perfectly correlated. The probability of having moldy slices in the package is $1 > p > 0$. If the first slice (B) taken from the loaf is moldy, what is the probability that the second one (A) is also moldy? The *conditional* probability that assets in tranche A (second adjacent slice is moldy) default, given that B defaults (first slice is moldy) is close to unity, $Pr(A|B) = 1$. In this case, the probability that the owners of tranche A lose is $Pr(AB) = Pr(A|B)\ Pr(B) = 1.Pr(B) = p$. The senior tranche is as risky as is the average of assets in the portfolio. Tranching/securitization can allow risk reduction only if there is no systemic risk. With systemic risk, where $Pr(A|B) = 1$ or close to it, the whole structure can collapse. If B defaults one is confident that A will default.

Since there were many assets in the CDO and no due diligence was performed, it was extremely difficult or impossible to evaluate the correlations or conditional probabilities of defaults among tranches. Because there was no organized exchange for the CDOs, their market prices were not transparent. No one knew what they were worth if sold on the market. This was the challenge for the Quants and rating agencies.

The "solution" was Li's copula/CDS model. One would know if they viewed "the apples in the basket" or "slices of bread" case as more relevant. In the "slices of bread" case, there was no point in correlating tranches. Only the probability p that the package will default is relevant.

The CDS rate was linked closely with house prices. Because the CDO boom was occurring at the same time as the housing prices were inflating, the CDS showed very little risk. In fact, there was a positive feedback between the CDS rate and housing prices. The rise in house prices lowered the CDS rate on the tranches and the packages of tranches increased in value. This increased the demand for mortgages by the securitizers, which in turn induced a greater supply of mortgages of any quality. Thereby housing prices rose further.

3.4 Major Premise of Economics/Finance: No Arbitrage Principle (NAP)

The Quants relied upon three models: the CAPM, the Black-Scholes-Merton Options pricing model (BSM) and the Efficient Market Hypothesis (EMH) to price securities and manage risk. The first two models are logical deductions from their premises. The BSM was a brilliant synthesis of mathematics and economics. Both the CAPM and BSM were based upon the major premise in economics and finance: the *No Arbitrage Principle* (NAP).

The NAP is the foundation of microeconomic household and firm optimization. A main principle in economics is that the gain per dollar spent on a consumer good should be the same for all goods consumed. Similarly in the theory of the firm, the gain per dollar spent on an input in the production process should be the same for all inputs used. The NAP conditions are expressed as follows. Let the consumer have a utility function over n-goods or services. The price of each good is known and fixed. The budget constraint is that the sum of expenditures over the n-goods equals the given budget. The NAP equation is that the marginal utility per dollar should be the same for each of the n-goods consumed. The marginal rate of substitution in utility (along the indifference curve or surface) should equal the relative given market prices (slope of budget line). Utility is thereby maximized subject to the budget constraint. If the relative marginal utilities are given, the relative prices are determined, and vice-versa.

In the theory of the firm, a known production function relates inputs to output. For a given cost, the inputs used should be such that their marginal product per dollar of input price be equal for each input used. Equivalently, the marginal cost to produce output (price input/marginal product) should be the same for all inputs used. If one knows the relative marginal productivities, then relative input prices are known and vice-versa. This no arbitrage condition NAP minimizes costs for a given output.

3.4.1 *CAPM Model*

The no arbitrage condition takes an analogous form in CAPM and Black-Scholes-Merton models in finance. The NAP principle in the CAPM was derived by Sharpe (1964) as follows. Let there be a market (m) portfolio of n assets. Each asset has a return and a risk/variance, and there are n^2 variances and covariances.

The market portfolio (designated by "m") has an expectation μ and a variance $\sigma_m{}^2$, based upon the variance/covariance matrix of the n-assets. There is also a safe asset with an interest rate of r. One can achieve a combination of expected returns and risk by combining the market portfolio whose return – risk is (μ, σ_m) with the safe asset whose return – risk is (r, 0). This linear combination is the capital market line, where the risk-return trade-off is:

$$dE/d\sigma_m = (\mu - r)/\sigma_m. \tag{3.1}$$

One can also achieve a risk – return by varying the composition of the market basket. Let there be a *portfolio* where fraction a is in asset i and fraction (1−a) is in the entire market basket of all n = 1,2,. . .i,. . .n-assets. The expected return is E and the variance is σ^2 for the portfolio (3.2).

$$E = aE_i + (1 - a)\mu \quad variance = \sigma^2 \tag{3.2}$$

As fraction *a* varies, both the expected return and the risk of the portfolio change. One obtains dE/da and dσ/da by varying the composition. Equation 3.3 is the risk-return trade-off obtained by varying the composition of the portfolio, *evaluated at the optimum portfolio*, when a = 0.

$$\{[(dE/da)/(d\sigma/da)]|a = 0\} \tag{3.3}$$

The NAP requires that the two risk-return trade-offs be equal. Equation 3.4 equates (3.1)–(3.3).

$$\{[dE/d\sigma_m] = (\mu - r)/\sigma_m\} = \{(dE/da)/(d\sigma/da)]|a = 0\} \tag{3.4}$$

From (3.4) one obtains the CAPM equation (3.5).

$$[E_i - r] = \beta_i[\mu - r], \quad \beta_i = \sigma_{im}/\sigma_m^2 \tag{3.5}$$

The CAPM provides a good measure of risk. Assets can only earn high average returns if they have high betas. Average returns are driven by beta because beta measures the extent that the addition of a small quantity of the asset to a diversified portfolio adds to the volatility of the portfolio. Beta is estimated by regressing the observed return of asset i upon the observed return on the total portfolio, each adjusted for r, the safe rate of return.

3.4.2 BSM Model

The BSM options pricing is based upon the NAP. Fischer Black derived the equation on the principle: Assets must be priced such that risk per unit of return is the same for all traded assets. Let there be a stock whose price S(t). The price change dS(t) is the sum of a drift plus a Brownian Motion (BM) term dw(t), where $E(dw(t)) = 0, E(dw(t))^2 = dt$. The drift coefficient μ and diffusion coefficient σ are constant.

$$dS(t) = \mu S(t)dt + S(t)\sigma dw(t) = drift\, S(t) + diffusion\, S(t). \tag{3.6}$$

There is a derivative whose value V is linked to the price of the stock S(t). There are several ways to derive and view the BSM equation and I follow the Fleming-Soner (2006, pp. 360–362) exposition within the NAP. One can buy the derivative at time t whose value V(t,S) depends upon the stock price at time t, or one can purchase for the same amount a portfolio of value X(t) = V(t,S), consisting of the stock and a bond yielding a safe return of r.

The change in the value of the portfolio dX(t), (3.7) is the sum of a drift term and a diffusion term. Similarly the change in the value of the derivative dV(t) is the sum of a drift and a diffusion term (3.8).

$$dX(t) = drift\, X(t) + diffusion\, X(t) \tag{3.7}$$

$$dV(t, S(t)) = drift\, V(t) + diffusion\, V(t) \tag{3.8}$$

The NAP states that, for the same risk, the change in the value of the portfolio dX should equal the change dV in the value of the derivative dX(t) = dV(t, S(t)). Use (3.7) and (3.8) to derive NAP equation (3.9).

$$drift\, X(t) + diffusion\, X(t) = drift\, V(t) + diffusion\, V(t) \tag{3.9}$$

The NAP requires that both assets having the same risk must have the same return. Equate drift X(t) to drift V(t) (3.10). This equalizes the return. Equate the two diffusion terms (3.11). This equalizes the risk. The portfolio composition at each time must be adjusted to provide an equal risk.

$$drift\, X = drift\, V \tag{3.10}$$

$$diffusion\, X = diffusion\, V \tag{3.11}$$

Solve the two equations for the ratio of the stock/portfolio and for the value of the derivative V. The BSM equation (3.12) for derivative pricing V(t,S(t)) is derived.

$$V_t\,(t, S) + rSV_s(t, s) + (1/2)\sigma^2 S^2 V_{ss}(t, s) = rV(t, S). \tag{3.12}$$

The main implications of BS are: (a) the drift term μ of the stock plays no role in the pricing. (b) The variance σ^2 plays a very important role.

3.4.3 The Efficient Market Hypothesis (EMH)

An important hypothesis used in the finance literature is the EMH Efficient Market Hypothesis. It is based upon the hypothesis that "Properly Anticipated Prices Fluctuate Randomly". However, its application to the world of finance with the derivatives rests upon vague and arbitrary assumptions. This was the fatal flaw. First I discuss these models and then evaluate their application to the pricing of derivatives.

The EMH is an empirical application of the hypothesis that "Properly Anticipated Prices Fluctuate Randomly" (Samuelson 1965). The latter is a set of mathematical propositions. The EMH has been applied to stock prices and foreign exchange rates with less than successful verification. I explain later in this chapter why the attempt to apply it pricing of CDOs, CDS and tranches failed and led to the bubble and its collapse.

The view underlying the EMH is that in an "informationally efficient" market price changes must be unforcastable. By "informationally efficient" one means that the market price fully incorporates the expectations and information of the market participants (Fama 1970). The mathematical structure of the argument below is based upon Feller, volume II (1966). It consists of three propositions, any one will imply the other two.

Let X(t) be the change in price P(t). Equation 3.13 states that the expectation of the change in price E[X(t)] = 0. This proposition is called an absolutely fair game.

$$E[X(t)] = 0. \tag{3.13}$$

It reflects the view that, in a competitive market with informed buyers and sellers trading at market determined prices, if one were sure that a price will rise, it will already have risen.

Price P(t) is the sum of previous price changes X(s), s < t. It can be expressed as the previous price P(t−1) plus the current price change X(t) (3.14).

$$P(t) = \sum(X(s)) = P(t-1) + X(t). \quad t \geq s \geq 0. \tag{3.14}$$

The expectation of the price E[P(t)|P(t−1), P(t−2)...], conditional upon its past history,

$$EP(t) = E\left[\sum\left(X(s)\right)\right] = P(t-1) + EX(t) = P(t-1) \tag{3.15}$$

is (3.15), because (3.13) states that the expectation of a price change E X(t) is zero. This proposition is the *martingale property*. The expectation of the next period's price is the current price.

The third proposition is called the *Impossibility of Systems*. It states that no strategy based upon the past history of prices can be profitable. No system can produce positive expected profits. Although a system may work at one time, that is a fluke and what works once is unlikely to work again.

Z(t) is cumulative profits. Let v(t) be a system, a formula, fixed in advance. It tells one when and how much to buy/sell at any time, based upon the past history of prices or any other variables. Then Z(t)

$$Z(t) = Z(t-1) + v(t)[P(t) - P(t-1)] = Z(t-1) + v(t)X(t). \tag{3.16}$$

is the sum of the previous cumulative profits Z(t−1) plus the profit from the last trade, based upon the fixed in advance system v(t).

The expectation of cumulative profits E[Z(t)] conditional upon the past prices is

$$E[Z(t)] = Z(t-1) \quad sinceE[X(t)] = 0 \tag{3.17}$$

It is equal to the previous profits, since the expected price change EX(t) = 0 from (3.13). Any sequence of decision functions converts martingale {P(t)} into martingale {Z(t)}.

Any one (3.13), (3.15), (3.17), which describes the hypothesis that "Properly Anticipated Prices Fluctuate Randomly", implies the other two. This is a mathematical proposition but it does not explain where the basic probability distributions underlying Expectation *E* come from. It does not explain what is "Information" nor does it explain how market participants act upon what they think is "information" or by "risk adjusting" their expectations.

The next section discusses how the market/Quants used these hypotheses in the world of CDOs, CDS, tranches and options, and how their misuse led to and aggravated the bubble.

3.4.4 The Quants and the Models

On the basis of the EMH and CAPM, Greenspan, the Fed and the finance profession believed that markets would be self-regulating through the activities of analysts and investors. In their view, government intervention weakens the more effective private regulation.

Securitization/tranching, the various layers of CDOs, and CDS, produced an environment where the EMH/CAPM lost relevance. These bundles of many mortgage based securities seemed to tailor risk for different investors. Securitization/tranching gave the illusion that one could practically eliminate risk from risky assets and led to very high leverage, as discussed in the section on the Atlas fund. In hindsight, it was difficult to understand how tranching the equity tranche – which consisted of poor prospects – into tranches could produce AAA ratings.

In reference to the LTCM collapse, Derman (p. 190) wrote: "It was a shock to realize that people whose great experience and knowledge straddled both the quantitative and trading worlds had, despite their sophistication, brought themselves into such a catastrophic state". The same could certainly be said about the financial collapse of 2007–2008. How can the latter be explained?

The Quants made several serious errors in modeling. First, there were no relevant "betas" to measure the risk of the CDOs. Second, the martingale property of the price is a special case of "The Principle that Properly Anticipated Prices Fluctuate Randomly", which does not apply to the house price index. Third, they ignored the "no free lunch" principle, discussed below, that the expected present value of an asset must be finite. Fourth, they assumed that there is a stable distribution function for house prices, from which they could derive the VaR value at risk. They assumed that the "free lunch" would continue, and invoked a "Black Swan" to justify their failures Fifth, they ignored the feedback in both directions between house prices, the price of a CDS, and the debt of households. Each of these is discussed in turn.

3.4.5 The CAPM

The financial market consisted of several stages. At one end were the mortgagors, households who borrowed against negligible collateral or ability to service the loans from income. Their loans were packaged or securitized into bundles by financial intermediaries who could not perform due diligence since they had no idea of the quality of the loans. In turn these packages were tranched and sold to institutional investors, who relied upon the rating agencies to evaluate risk. How could the risk be evaluated? The CAPM states that it is measured by the beta of the tranche (3.5) repeated.

$$[E_j - r] = \beta_j[\mu - r], \quad \beta_j = \sigma_{jm}/\sigma_m^2$$

To calculate the beta of tranche j one must know, the expected return μ on the larger portfolio containing tranche j, the variance-covariance matrix of all of the tranches and assets in the larger portfolio, the covariance σ_{jm} to between the return on tranche j and the return on the larger portfolio. These quantities could not be calculated. There were thousands of mortgages in each tranche, most of them of dubious value. No due diligence was performed. Their market prices were unknown so that one could not calculate the E_j the expected return on the tranche. In addition, what was the market portfolio containing tranche j whose mean was μ, whose variance was σ_m^2 and covariance was σ_{jm} with tranche j?

3.4.6 Credit Default Swaps, EMH and the House Price Index

Since there are many assets in the CDO and no due diligence was performed, it was extremely difficult or impossible to evaluate the correlations or conditional probabilities of defaults among tranches. There was no organized exchange for the CDOs. Their market prices were not transparent. No one knew what they were worth if sold on the market and internal accounting rules allowed Mark-to-Market values to be set by internal risk models rather than by verifiable quotes from transaction. This was a challenge for the Quants and rating agencies.

The "solution" was Li's copula/CDS model. His model could assign correlations between the tranches by measuring the Credit Default Swap rate (CDS) linked to the underlying security. The CDS of the tranches supplied a single variable that incorporated the market's assessment of how the tranche will perform. The CDS rate was linked closely with the prices of the mortgages and CDOs, which in turn was based upon the house price index. Then one could price the CDOs/tranches based upon the CDS rate. The important question was how to estimate the expected house price, in order to price the CDS.

EMH equation (3.15) states that the expected price EP(t + dt) is equal to the present price P(t), the martingale property. This is only true, however, in special cases and is not implied by the NAP. For example, the spot price of a harvest commodity is not a martingale, whereas the futures price is a martingale (Stein 1986; Samuelson 1957).

Inventories rise as the harvest comes in. The balance of short hedging by those carrying inventories lowers the spot price below the price expected at a later date, due to a risk premium and storage costs. As inventories decline, short hedging declines and the spot price rises towards the expected price. Thus P(t) is less than EP(t + dt) and its expected change E[dP(t)] is not zero.

Similarly, before the harvest there is a balance of long hedging, by those who want to protect themselves from later purchases at higher prices. The spot price P(t) rises above the expected price, due to a risk premium and very low inventories. As the harvest comes in, there is a decline in long hedging, and the spot price declines. Thereby P(t) is greater than EP(t + dt) and its expected change E[dP(t)] is not zero. The NAP is not violated. In both cases however, the futures price is tied to EP(t + dt) and is a martingale.

In the case of the CDS, the variable to consider is the index of house price P(t). Let the house price follow a random walk with drift, equation

$$dP(t) = \mu P(t)dt + P(t)\sigma dw(t) = drift + diffusion, E[dw] = 0, \ \ E\left[dw^2\right] = dt. \tag{3.18}$$

The drift coefficient, the trend, is μ and the diffusion is $\sigma \, dw(t)$. The CDS rate is based upon the expected house price index. Solve (3.18) for P(t), using the Ito equation to derive (3.19). Take the expectation and derive (3.20).

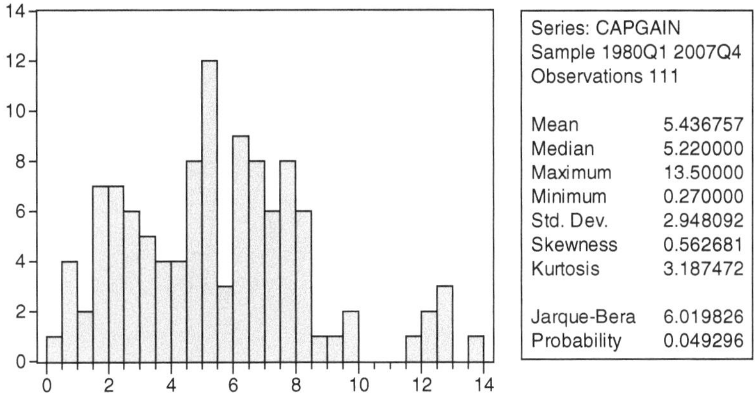

Fig. 3.4 Histogram and statistics of CAPGAINS = Housing Price Appreciation HPA, the change from previous four-quarter appreciation of US housing prices, percent/year, on horizontal axis. Frequency is on the vertical axis. Period 1980q1–2007q4 (*Source*: Office of Federal Housing Price Oversight)

$$P(t) = P(0)exp\{\int[(\mu - (1/2)\sigma^2]dt + \int \sigma dw\}\ t > 0. \tag{3.19}$$

$$EP(t + dt) = P(t)exp\left\{\int \mu\, d\tau\right\},\ t + dt > \tau > t. \tag{3.20}$$

The expected house price that determines the CDS rate is (3.20). *This is a rational expectations price*. Only if $\mu = 0$, there is no drift in the house price index, would the house price and the underlying CDS be a martingale.

The CDS rate was then used as a measure of risk of the tranche. One must have estimates of μ to form an estimate of the expected house price that determines the CDS rate and price/risk of the tranche. Since the CDO boom was occurring at the same time as the housing prices were inflating, the CDS were showing very little risk according to this view.

To estimate the trend μ or drift the Quants would have to estimate the distribution of house prices. Then they could use a VaR to determine the probability of a decline in house prices. There were several possibilities. (1) Assume a stable distribution of house prices, based upon historical data; (2) Assume an unstable, a changing, distribution; (3) Assume a stable Pareto-Lévy distribution which has a fat tail; (4) Assume a jump diffusion process.

Figure 3.4 is a histogram of the distribution of [P(t) – P(t−1)]/P(t−1) the percentage change in house prices, the capital gains – over the period 1980q1–2007q4 prior to the crisis. These data could be used to infer the drift and risk parameters underlying the expected price. A VaR could then be derived.

The histogram shows several factors. (1) Estimates of trend $\mu = 5.4\%$ pa and of risk $\sigma = 2.9\%$ pa. (2) The null hypothesis is that distribution is normal. The probability is about 5% that the Jarque-Bera statistic exceeds the observed

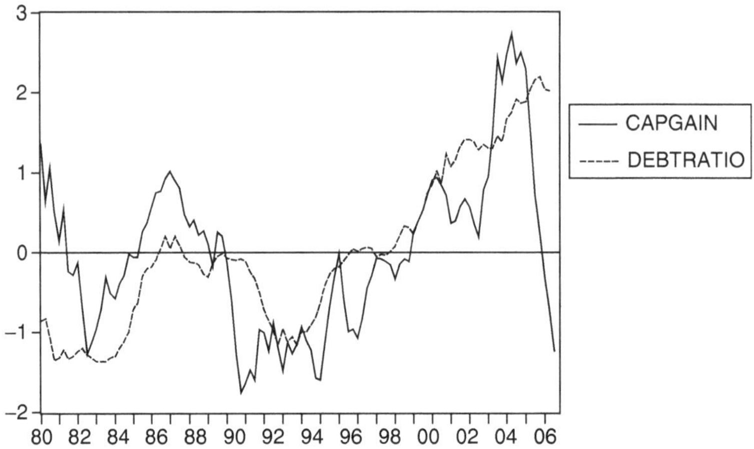

Fig. 3.5 Capital gains [P(t) – P(t−1)]/P(t−1) = CAPGAIN; DEBTRATIO = household debt/disposable income (Variables are normalized to a mean = 0 and a standard deviation = 1)

value under the null hypothesis. Small probability value leads to the rejection of the null/normality. (3) There were no sub-periods of falling house prices. (4) There is a positive tail. These were the "bubble years".

Figure 3.5 is a time series of capital gains [P(t) – P(t−1)]/P(t−1) CAPGAIN and the debt/disposable income of households DEBTRATIO, over the period 1980q1–2007q4. A major error that the Quants made was to assume that the distribution described by the histogram Fig. 3.4 is stable. They assumed that these capital gains could continue while the debt/income ratio of households was rising steadily relative to its longer run mean value. The distribution of the capital gains was very different in the period 2000–2011.

The capital gains from 2004 to 2006 reached 2.9 standard deviations above the mean for the entire period 1980q1–2007q4. The mortgage interest rates were between 5% and 7% per annum as seen in Fig. 3.3. Thus during the 2004–2006 period the capital gains exceeded the mean interest rate. If an estimate of the drift term μ or expected capital gain exceeds the mean interest rate r, then the expected present value PV of the asset $PV(t) = P(0) \exp[(\mu - r)t]$ diverges as time increases. That makes no sense.

Similarly, the Quants ignored the fact that many mortgagors were enjoying a "free lunch". The mortgagor had a mortgage of M(1), say equal to the value of the house, at interest rate r(t) for one "period". At the end of the period, the debt was $M(1)(1 + r(t))$. Insofar as housing prices rose by $\pi(t)$ percent over the period, the value of the house was $M(2) = M(1)(1 + \pi(t))$. Insofar as the capital gain $\pi(t) > r(t)$, the mortgagor refinanced and took out a mortgage equal to M(2), against equity. He repaid the loan and used the proceeds $M(1)(\pi(t) - r(t))$ for expenditure. This procedure was continued/repeated as long as the capital gain $\pi(t)$ exceeded i(t) the interest rate. The difference $M(t)(\pi(t) - r(t))$ was a free lunch that could not continue. As soon as the capital gain fell below the interest rate, the value of the house would be less than the debt. Delinquency and foreclosure would follow. This indeed occurred.

3.5 When Has the Drift Changed?

When the collapse occurred some Quants tried to adjust the stochastic process by assuming a jump diffusion process. To a constant diffusion of prices they assumed that there was a small probability that the price might take a sizeable jump. This ad hoc procedure is not useful. How can one estimate this small probability, especially since there were very few observations of a crisis? Figure 3.4 shows that there were no price declines from 1980 until the 2007 crisis. From the time series and distribution of prices, there is no information of when or by how much the jump will occur. There is no early warning signal of a jump or when the drift has changed.

A much more rigorous and elegant approach to the problem "When has the drift changed?" is the article by Blanchet-Scalliet et al. (Bl-S 2007). An investor divides his wealth between a safe asset and a stock – the risky asset. The price of the stock has a drift and a diffusion. Their paper derives an optimal stopping or switching rule, given that the drift will change at some unknown time. The problem is to find an "alarm signal" for when to take some action.

To relate their work to this chapter let the price of the risky asset be the house price index P(t). The risky asset is a CDO or CDS whose price is linked to the house price P(t). The house price P_t evolves according to (3.21), (3.22), (3.23), (3.24), and (3.25), where μ_t is the drift and σ dw is the diffusion. The past/current drift is μ_1 and we anticipate that it will change to μ_2 at an unknown later date from the present. The expected drift is $E\mu_t$.

$$dP_t = P_t(\mu_t\, dt + \sigma dw) \quad P = P(0) \quad dw \sim N(0, dt) \tag{3.21}$$

$$d\, ln\, P_t = (\mu_t - \sigma^2/2)dt + \sigma dw_t \tag{3.22}$$

$$ln\, P_t = ln\, P + \int(\mu_s - (1/2)\sigma^2)ds + \sigma w(t) \qquad w(t) = \int^t dw_s ds \tag{3.23}$$

$$E\, ln\, P_t = ln\, P + \int^t (E\mu_s - (1/2)\sigma^2)ds \tag{3.24}$$

$$var\, ln\; P_t = \sigma^2\, t \tag{3.25}$$

Bl-S postulate the *unconditional* probability, that there is no change in the drift, is (3.26a), and the *unconditional* probability of a change is (3.26b). If Pr $(\mu_t = \mu_1) = 0.5$, then the half-life in state μ_1 is (3.26c). Thus $\lambda > 0$ is inversely related to the half – life in state 1.

$$Pr\,(\mu_t = e^{-\lambda t}) \quad t \geq 0 \tag{3.26a}$$

$$Pr\,(\mu_t = \mu_2) = 1 - e^{-\lambda t}. \tag{3.26b}$$

$$ln\,(0.5)/\lambda = 0.69/\lambda = half - life \tag{3.26c}$$

From (3.21, 3.22, 3.23, 3.24, 3.25, 3.26) how should one determine if drift μ has changed? This is an important question since it determines the optimal ratio of risky assets/wealth and the implied optimum growth rate of wealth.

They contrast three approaches. In one, the optimization is based upon a misspecified model, for example where future changes are ignored. The second is empirical. A moving average of prices is used to estimate when the drift has changed. Chartist trading rules are just based upon the history of the asset price. The third are two model-detect strategies.

Their model dependent optimization procedure is directly related to the question when has the drift changed? This challenge is illustrated by Fig. 3.5 above concerning the capital gain. I sketch the Bl-S approach.

Utility function of wealth ln X(t), implies considerable risk aversion. The aim is to select the proportion π_t in the risky asset to maximize the expected logarithm of wealth ln X(T) to time T > 0. That is, π_t maximizes (3.27).

$$G\,(X) = max\,E\,ln\,X_T/X, \quad X = X(0). \tag{3.27}$$

X = wealth = value of risky asset + value of safe asset

Their constraint is that $1 > \pi > 0$. Assume that $\mu_1 > r > \mu_2$. The drift starts above the safe rate and then will fall below it at some unknown time. The question is when one should reduce or even get out of the risky asset. When will the bubble burst? The expected drift times the fraction of wealth in the stock is $\pi_t E\mu_t$.

Using the stochastic calculus Ito equation they derive (3.28) from (3.27), (3.22) and an equation for wealth, a linear combination of the stock and the safe asset.

$$ElnX_T/X = \int^T [\pi_t E\mu_t - \pi_t{}^2\sigma^2/2]dt + rT \tag{3.28}$$

Hence the optimal proportion π^*_t to be held in the risky asset is (3.29).

$$\pi_t^* = argmax[\pi_t E\mu_t - \pi_t{}^2\sigma^2/2] \tag{3.29}$$

This ratio maximizes the expected utility, the expected logarithm of wealth at time T.

Their goal is to find a stopping rule that detects at what instant τ the drift of the house price index has changed. Since the investor observes the house price P(t), but cannot observe the drift $\mu(t)$, this is a partially observed optimal stopping problem.

A technique using nonlinear filtering formulas to reduce this partially observed problem to a completely observed problem was first used by Shiryayev and has now become standard. Bl-S et al compute the wealth of the trader who uses one of the two Model and Detect Strategies, relative to the Moving Average atheoretical approach. They ask. *Is it better to invest according to mathematical Model and Detect strategy based upon a mis-specified model, or according to a strategy which is model free?*

The main mathematical tool used to obtain these two stopping rules is the process F_t, the *conditional* probability that the (unknown) change point appeared before the running time t. For each procedure, the trader decides to invest his wealth in the risky asset when F_t differs from a given quantity defined in their paper.

The Bl-S equation (3.30) is the *conditional expectation* of the drift μ^*_t. Variable F_t, is the conditional probability that the change in the mean rate of return from has occurred by time t, given the observed house prices up to that time. If $F = 1$ then the jump down from μ_1 to μ_2 has occurred. If $F = 0$, the jump has not yet occurred. The mean rate of return μ_t is a hidden state with two possibilities, μ_1 or μ_2, Only one change in the state is allowed at an random exponentially distributed time τ. This is different from an *unconditional* expectation based upon (3.26a), (3.26b). Using (3.30)

$$\mu_t^* = [\mu_1 - \sigma^2/2) + (\mu_2 - \mu_1)F_t] \tag{3.30}$$

in (3.29), they determine the optimal ratio of the risky asset/wealth π^*_t which maximizes expected logarithm utility given the stochastic process on the drift.

The evolution of F_t, the conditional probability that change has already occurred, is based upon a likelihood ratio and an *innovation* process. It depends upon both the observed house prices and the conditional probabilities F_s, for $s \leq t$. The details are in their article. For my purposes, the derived conditional probability has the following characteristics [Bl-S, eqs. A6, A7].

Start with a drift μ_1. One has a prior and a recent sample. The forecasting error compares the current price realization ln P(t) with what is expected from the prior drift $[\mu_1 - \sigma^2/2]t$. The difference is used to obtain the probability F_t that the change in drift from μ_1 to μ_2 has already occurred.

They conclude (p. 1,366): "We can now address our main question: Is it better to invest according to a mathematical Model and Detect strategy based upon a mis-specified model or according to a strategy which is model-free? Due to the analytical complexity of all the explicit formulae that we have obtained for the various expected utilities of terminal wealth, we have not as yet succeeded to find a mathematical answer to this question... We therefore present numerical results obtained from Monte Carlo simulations to illustrate our comparisons."

Bl-S simulate the models and conclude (p. 1,367) "Our numerical study suggests that there is no universal solution to the problem of parameter mis-specification. It seems that when the drifts are high in absolute terms and in particular, when the upward drift is high, the performance of the Model and Detect strategies can be quite robust and superior to the one of the chartist trading strategy. However, their performance deteriorates rapidly when λ is strongly mis-specified and/or when the upward drift is not very high. Since the second drift is in fact the hardest to estimate due to the fact that we lack a priori information, we recommend caution before asserting that the Model and Detect strategies are superior to the technical trading rule. Indeed the Model and Detect strategies only offer a clear comparative advantage over the chartist trading rule in the presence of strong expected future trends".

3.6 Conclusion: Errors of the Quants

According to the EMH, it is impossible to know when a bubble is occurring because current prices reflect all publicly available information. Only in retrospect, when the bubble has collapsed, can one infer that prices did not reflect "the fundamentals". This was precisely Greenspan's view (Chap. 2): the crisis was unpredictable. The theme of this book, in chapters below, is that Greenspan was in error.

The Quants/market made three interrelated errors. First, they based the expected house price upon the current price and drift according to (3.20). When the drift $\mu = 0$, the expected price was just the current price (EMH). A second closely related error is that they assumed that the distribution of capital gains is constant, so they could derive a VaR on that basis. The mistake was that the Quants ignored publicly available information in forming expectations, as discussed in Chap. 2. The ultimate determinant of the values of CDOs was the ability of the mortgagors to service their debts. The income to the owners of the CDOs and their values comes from the flow of income through the tranches starting with the mortgagors. It was public information that the mortgages were of most dubious value, no due diligence was performed in the ratings and the debt/income of the mortgagors – DEBTRATIO in Fig. 3.5 was rising. Hence it was improbable that the distribution of house prices would remain constant. This systemic risk was ignored by the Quants who just focused upon current prices.

A third major error of the Quants was to assume the independence of the returns to the tranches, the "apples in the basket" model, instead of the correlations in the "slices of bread in a loaf" model.

References

Blanchet-Scalliet, C., A. Diop, R. Gibson, D. Talay, and E. Tancré. 2007. Technical analysis compared to mathematical models based methods under parameter mis-specification. *Journal of Banking and Finance* 31(5): 1351–1373.

Cheng, Ren. 2009. *Some reflections on the current state of financial mathematics*, Graduate International Colloquia Grant, Challenges of Mathematical Modeling and Numerical

Computation in Finance, Division Applied Mathematics Brown University. <www.dam.brown.edu./colloquium/DAM_colloquium/Schedule.html>.

Congressional Oversight Panel (COP). 2009. Special Report on Regulatory Reform, Washington, DC.

Derman, Emanuel. 2004. *My life as a quant*. New Jersey: Wiley.

Fama, Eugene. 1970. Efficient capital markets: A review of theory and empirical work. *Journal of Finance* 25(2): 383–417.

Feller, William. 1966. *An introduction to probability theory and its applications II*. New York: Wiley.

Financial Crisis Inquiry Report (FCIC). 2011. Perseus Group. New York.

Fleming, Wendell, and Mete Soner. 2006. *Controlled Markov processes and viscosity solutions*. New York: Springer.

Patterson, Scott. 2010. *The quants*. New York: Crown Business.

Samuelson, Paul. 1957. *Intertemporal price equilibrium: A prologue to the theory of speculation*. Reprinted in Collected Papers (1966) II.

Samuelson, Paul. 1965. *Proof that properly anticipated prices fluctuate randomly*. Reprinted in Collected Papers (1972), III.

Sharpe, W.F. 1964. Capital asset prices: A theory of market equilibrium under conditions of risk. *Journal of Finance* 19(3): 425–442.

Stein, Jerome L. 1986. *The economics of futures markets*. Oxford: Blackwell.

Yang, Jichuan. 2009. *The making of a beautiful thing, or a financial weapon of mass destruction*, Graduate International Colloquia Grant, Challenges of Mathematical Modeling and Numerical Computation in Finance, Division Applied Mathematics Brown University. <www.dam.brown.edu./colloquium/DAM_colloquium/Schedule.html>.

Chapter 4
Philosophy of Stochastic Optimal Control Analysis (SOC)

Abstract I discuss the philosophy of the stochastic optimal control techniques used in later chapters. The net worth of the real estate sector in Chap. 5, and of AIG in Chap. 6, evolve dynamically. In the first case, debt is incurred in period t to purchase assets whose return is uncertain, and must be repaid in period t + 1 at an uncertain interest rate. In the second case, insurance is sold in period t and the claims in period t + 1 are uncertain. What is the optimal debt in the first case and what are the optimal insurance liabilities in the second case? I discuss the strengths and limitations of alternative criterion functions, what should the firm or industry maximize? How should risk aversion be taken into account? Then I discuss the modeling of reasonable stochastic processes of the uncertain variables. Given the criterion function, each stochastic process implies a different quantitative, but similar qualitative, optimum debt/net worth or insurance liabilities/net worth. Using SOC I derive quantitative measures of an optimal and an excessive leverage, or what is an excessive risk that increases the probability of a crisis? The optimal capital requirement or leverage balances risk against expected growth and return.

4.1 Why Use Stochastic Optimal Control?

Neither the markets, the IMF nor the Central Banks anticipated the US financial crises until it was too late. The collapse of the housing market led to solvency problems – where debt exceeded assets – and liquidity problems, where key firms in the shadow banking system could not obtain sufficient liquid resources to service their debts. In many cases the failures led to bailouts, discussed in subsequent chapters. The basic question is: What is an "excessive" debt ratio for two key sectors in the crisis: the real estate sector and a large insurer such as AIG, that is likely to lead to a crisis? I explain why SOC is a valuable tool to derive theoretically based, not empirical ad hoc, measures of an excess debt that will be an Early Warning Signals (EWS) of a debt crisis.

J.L. Stein, *Stochastic Optimal Control and the U.S. Financial Debt Crisis*,
DOI 10.1007/978-1-4614-3079-7_4, © Springer Science+Business Media New York 2012

Net worth is assets less debt. The crucial variable is leverage or the debt/net worth ratio. Leverage = assets/net worth = 1 + debt/net worth. The growth of net worth is affected by leverage. An increase in debt at time t to finance the purchase of assets increases the growth of net worth by the return on investment. The return on investment has two components. The first is the *productivity of assets* and the second is the *capital gain* on the assets. The resulting higher debt decreases the subsequent growth of net worth by the associated interest payments and servicing of the debt. As a result an increase in leverage will increase expected growth of net worth if the return on investment exceeds the interest rate on the higher debt. The productivity of assets is observed, but the future capital gain and the interest rates are unknown when the investment decision is made.

The basic equation for the growth of net worth is crucial to understanding optimal risk management and evaluation of desirable policy. Two cases are discussed in detail in the chapters below. In Chap. 5, the focus is upon the real estate sector and the US financial crisis. In Chap. 6, the focus is upon AIG which sold Credit Default Swaps (CDS) – a form of insurance – against the financial derivatives that were based upon debt from the real estate market. In both cases, I use the SOC approach.

The first case concerns the real estate market. The mortgagors/households borrow from banks, which then package the mortgages and sell these packages to investors. Focus upon the net worth of the mortgagors/real estate industry whose interest payments become income to the buyers of mortgages and their derivatives. The purchase of mortgages assets, which are investments in housing, implies a rise in the ratio of debt/net worth of the real estate sector. The income of the financial structure and the shadow banking system ultimately was derived from the ability of the mortgagors to service their debts. The increase in debt would raise the growth of net worth of the real estate sector if the capital gain on the assets – the appreciation of house prices – plus the rate of return on the assets exceed the interest rate (4.1).

$$\begin{aligned} \textit{Change in net worth} = {} & [\textit{assets}(\textit{capital gain} + \textit{productivity of assets})] \\ & - \textit{interest rate}(\textit{debt}) - (\textit{consumption or dividends}) \end{aligned} \tag{4.1}$$

The capital gains and the interest rates are described in Chap. 3. The huge capital gains in housing and low interest rates, during the period 2002–2006 led to a rise in the ratio debt/net worth of the real estate sector. The first term in brackets was large so the private housing sector increased its debt directly to banks and indirectly to foreign investors. The investment in housing seemed to be profitable because the debt could be refinanced/repaid from the recent capital gains – not from the productivity of assets/the "marginal product of capital". Those capital gains were not sustainable. The reason is that insofar as the capital gain exceeds the appropriate interest rate, the present value of the asset diverges to infinity.

There is a dynamic process. Borrowing to purchase assets in period t has a return. The value of assets changes, resulting from the capital gain (or loss). At time (t + dt) the unit has a greater debt that must be serviced. The higher debt raises the term [(interest rate) (debt)] in period (t + dt), which tends to lower the growth of net worth. Actions taken at time t have consequences at time (t + dt). The risk is that with the higher debt ratios, there would be a period when the capital gains fell below the interest rate – such as occurred in 2007–2008. In fact the capital gain term and interest rate term are negatively correlated. When interest rates declined, house prices rose – the boom/bubble. When interest rates rose, capital gains declined. When the growth of housing prices declined, then the large negative term – the return less the interest rate – is multiplied by a large debt ratio and the net worth of the housing industry vanishes. The housing sector defaults on their loans. Bank failures followed the collapse of the housing market. The government then intervened to avert the collapse of the financial system. The question is what is an optimal debt/net worth ratio at any time? How should one judge if the debt and risk were too large? The techniques of SOC are used in Chap. 5 to answer this question.

The second case concerning AIG is discussed in Chap. 6. The AIG Financial Products AIGFP subsidiary of AIG sold credit default swaps (CDSs) involving three parties. Party 1 the obligor sells the "reference securities" to party 2. The latter are security houses, hedge funds that purchase protection from party 3, who are banks and insurance companies, which issue the protection. In many cases municipalities are required to carry insurance in order to market their bonds, or to obtain a high credit rating. The CDSs are privately negotiated contracts that function in a similar manner to insurance contacts, but their payoff structure is closer to a put option. The CDS liability requires that the insurer pay, or put up more collateral, if the market value of the securities insured falls below the notional amount insured. Surplus is the difference between assets and liabilities less initial contributed capital. The change in surplus is the sum of several components (4.2).

$$\begin{aligned} \textit{Change in surplus} &= (\textit{premium rate})\ \textit{liabilities} + (\textit{return} + \textit{capital gain})\ \textit{assets} \\ &\quad - (\textit{claims on the company related to liabilities}) \end{aligned} \tag{4.2}$$

The insurer incurs a liability at time t, and receives a premium for the amount insured. This increases assets and surplus at time t. The capital gain (or loss) and claims at time t + dt are unknown at time t when the insurance is sold. The surplus at time t + dt rises as a result of the return, equal to the productivity plus the capital gain on assets. Surplus declines by the amount of claims at time t + dt against liabilities incurred earlier.

The stochastic variables are the capital gains on assets and claims against the insurer. They are unknown at time t when the liabilities are incurred. The control variable of the insurance company is its liabilities, the insurance policies sold such as CDS. The question is: what is the optimum amount of insurance liabilities? When should the firm and/or regulators decide that liabilities are too large?

In (4.1) the control variable is the debt of the real estate sector and in (4.2) it is the liabilities of the insurer. The capital gains, interest rate or claims in the cases above are subject to imperfectly known disturbances that may be taken as random. Stochastic optimal control theory attempts to deal with models in which random disturbances are very important. I deal with the case where the controller knows the state of the system at each time t. In deterministic systems, if one knows the initial conditions, the model and controls, then the paths of the variables can be predicted perfectly. For stochastic systems, it is completely different. There are many paths that the system may follow from the controls and initial data. In the stochastic case, the best system performance depends on the information available to the controller at each time t. The method of dynamic programming is generally used. The basic references for techniques used in this approach are: Fleming–Rishel, Chaps. V and VI, Fleming–Soner, Fleming–Stein (2004) and Stein (2005).

4.2 Research Strategy

My research strategy parallels the views of Fischer Black and Emanuel Derman (2004, ch. 16). I paraphrase/quote their views. Certain economic quantities are so hard to estimate that Black called them "unobservables". One unobservable is expected return. So much of finance deals with this quantity unquestioningly. Yet our estimates of expected return are so poor, that Black said that they are almost laughable.

The Value at Risk VaR uses statistical distributions to assign probabilities of the firm's value into the future, and statistics are inevitably based upon the past. But VaR is too simplistic. The statistical distributions are not necessarily stationary. The past does not repeat itself. People learn from experience, which subsequently changes the distributions. We are ignorant of the true probabilities – the extreme tails of price distributions.

Black described the approach of the finance profession as follows: "In the end a theory is accepted not because it is confirmed by conventional empirical tests, but because researchers persuade one another that the theory is correct and relevant". This "group think" behavior led to serious consequences as described in Chaps. 2 and 3.

The question is what is the "right model" to analyze problems posed in Sect. 4.1 above? The Quants are physicists, mathematicians and computer scientists who do "financial engineering". They apply models from physics to the world of finance. Derman has moved between the world of Wall St. Quants and academia. He wrote that models in finance differ from those in physics. "Models are only models, not the thing in itself. We cannot therefore expect them to be truly right. Models are better regarded as a collection of parallel thought universes you can explore. Each universe should be consistent, but the actual financial and human world, unlike the world of matter, is going to be infinitely more complex than any model we make of it. We are always trying to shoehorn the real world into one of the models to see how useful approximation it is".

His strategy, which I follow, can be summarized. One must ask: Does the model give you a set of plausible variables to describe the world, and a set of relationships between them that permits its analysis and investigation. The real question is how useful is the theory? "The right way to engage with a model is, like a fiction reader or a really great pretender, to temporarily suspend disbelief, and then push it as far as possible".

For these reasons, I consider a set of plausible models and derive the optimal controls, debt ratio or liability ratio, using SOC. Call the optimal ratio f*(t), which varies quantitatively among models. I then consider an upper bound on the optimal control for both models and call it f**(t). I compare the actual debt ratio or liability ratio, called f(t), with this upper bound. Thereby, I derive a measure of "excess debt" Ψ(t) = [f(t) − f**(t)], the actual less upper bound optimal. I explain that the probability of a debt crisis is directly related to Ψ(t) the excess debt. On the basis of this measure of excess debt, I derive early warning signals of the US financial crisis of 2008 and the debt crises of the 1980s.

4.3 Modeling the Uncertainty, the Stochastic Variables

In general, focus upon the change in net worth (4.1). It can be written as (4.3).

$$dX(t) = X(t)[(1 + f(t))\,(dP(t)/P(t) + \beta(t)dt) - i(t)f(t) - c\,dt] \qquad (4.3)$$

X(t) = Net worth, f(t) = debt/net worth = L(t)/X(t), dP(t)/P(t) = capital gain or loss, i(t) = interest rate, (1 + f(t)) = assets/net worth, β(t) = productivity of capital. C(t)/X(t) = c (t) = consumption/net worth = c is taken as given.

The sources of uncertainty are dP(t)/P(t) the capital gain or loss, and i(t) the interest rate. How should the stochastic variables be modeled? I consider in the chapters below two models that differ in their assumptions about the stochastic processes.

In Model I, assume that the price grows at a trend rate and there is a deviation from the trend. The deviation from trend is a stochastic variable that has two parts. The first part converges to zero and the second is random with a mean of zero and a positive finite variance. This deviation is an Ornstein Uhlenbeck ergodic mean reverting process. This implies a stochastic differential equation for the capital gains term. The interest rate has two parts, a constant mean plus a diffusion or random term, with a zero mean and finite variance. The capital gains term and interest rate are negatively correlated. Declines in interest rates lead to capital gains, and rises in interest rates lead to capital losses.

Since the growth of net worth equation (4.3) will differ according to the model assumed, the derived optimal debt ratio will differ in the two models. I examine the empirical data to decide which assumption is to be preferred. However, the tests do not allow one to reject either hypothesis.

Model II considers the return capital, the capital gain plus the productivity of capital, b(t) to be a stochastic variable. There are two parts to the return b(t). One is the mean $b\,dt$ and the second is a diffusion $\sigma_b\, dw_b$. The interest rate has a mean i dt and a diffusion $\sigma_i\, dw_i$ the latter is a random variable with a zero expectation.

$$b(t) = dP(t)/P(t) + \beta(t)dt = b\,dt + \sigma_b dw_b$$
$$i(t) = i\,dt + \sigma_i dw_i \tag{4.4}$$

Many reasonable models have the form of stochastic differential equation (4.5) for the change in net worth. The model based upon equations (4.4) is a special case.

$$dX(t) = A(t, X(t))dt + v(t)dt, \tag{4.5}$$

where X(t), v(t) are scalar valued. The mean is the first term and the term v(t)dt represents the disturbances. The finance and mathematics literature assume that the disturbance has the following properties. (1) v(t) is stationary with independent increments, (2) The increment v(t) − v(r), t > r, is Gaussian with mean of zero and a variance of $\sigma^2(t-r)$, (3) v(t) is Gaussian with a zero mean. Condition (2) follows from the others. In that case, the stochastic differential equation is (4.6)

$$dX(t) = A(t,\, X(t))\,dt + \sigma\,dw(t) \tag{4.6}$$

where the term w(t) is a Brownian motion and dw/dt is *formally* white noise. The expectation is zero E(dw) = 0, and the variance is σ^2 dt.

In fact, assumptions (1) and (2) do not accurately describe the empirical data. The disturbance term v(t)dt is generally serially correlated and is not normal. However, one may consider v(t)dt as wideband noise (Fleming–Rishel, p. 126). That is let v(t) be some stationary process with a zero mean and known variance-covariance R(r) = E[v(t), v(t + r)]. If autocorrelation R(r) is nearly 0 except in a small interval near r = 0, then v is called wide band noise. White noise corresponds to the case where R is a constant a times a Dirac delta function. Then v(t)dt is replaced by σ dw, where σ is a *constant* matrix. Then the corresponding diffusion leads to an approximation of solution to (4.5).

There are other realistic possibilities for modeling v(t), but they involve mathematical complexities and do not lead to (4.6) that characterizes the "Merton type" models. My strategy is to work with models I and II that lead to a stochastic differential equation (4.6) and allow for model uncertainty in deriving the optimal debt ratio.

The next challenge is how to model and estimate the mean term or drift in (4.6), E [dX(t)] = *A(t, X(t)) dt*. The Quants assumed that the distribution of the capital gains was constant, so they could obtain estimates of the drift. They had to use historical data, say 1980–2007. Ex-post, the basic distribution of capital gains changed drastically from 2000 to 2010 as shown in chapters below. How then should one know if the drift or distribution has changed? The issues of a time varying drift and detection of its change were discussed in Chap. 3.

An important constraint must be introduced. The Quants and market assumed a drift of the capital gain based upon recent data. This was a period of rapidly rising house prices where the mean was considerably higher than the mean rate of interest.

Mortgagors refinanced their homes and incurred a debt equal to the current high house prices. The latter exceeded the value of the initial mortgage. They consumed the monetary difference between the value of the new loan and the value of the old loan. In the next period, insofar as the house price appreciated at a greater rate than the interest rate, they were able to repeat the process: refinance and consume the difference between the value of the home and the new debt. This was a free lunch. The mortgagor may have been a NINJA (no income, no job, no assets) and the loan originator did not worry for two reasons. First, the equity in the home was rising due to the capital gain. Second, the originator sold the mortgages that were then packaged and sold to investors. Free lunches cannot continue. I therefore constrain the optimization: the drift of the capital gain should not exceed the mean rate of interest.

4.4 Criterion Function

Chapters 2 and 3 explained that the finance industry has been poor at risk management and that the regulators have been ineffective. I assume that there is a hypothetical or ideal optimizer. In the financial market case it is the real estate industry. In the insurance case it is AIG. In each case the hypothetical optimizer selects a debt ratio that maximizes the expectation of a concave function of net worth Max EU(X(T)) at a terminal date T, subject to dynamic stochastic processes. This is a benchmark for optimality.

The criterion function should satisfy several requirements. (1) Losses are penalized more than equal gains are rewarded. This requirement seems to have been ignored or reversed in the finance industry/Wall St. prior to the crisis. What the market seemed to optimize differed from the requirements of an "optimal" solution. (2) Very low values of X(T) should be very severely penalized. This too was ignored due to the moral hazard where major firms anticipated that a friendly Treasury/Federal Reserve would bail out "too big to fail" firms. In modeling, the distribution of the stochastic variable in (4.6) is assumed to be normal, but in reality it is not. There may be a fat tail, where there is a low probability of very large losses. Large losses may not be likely, but they should get very heavy weights in the criterion/utility function.

Utility *U(X-Y)* is (4.7), using a second order approximation around U(X) for small values of stochastic variable Y. It has a zero mean and a variance of σ^2

$$U(X-Y) = U(X) - U'(X)Y + (1/2)U''(X)Y^2 \tag{4.7}$$

$$\mathrm{E(Y)} = 0. \tag{4.8}$$

$$\mathrm{E(Y^2)} = \sigma^2. \tag{4.9}$$

Table 4.1 Alternative criteria/utility functions

Criterion function U(X(T))	Lim X(T) as X = > 0	Risk aversion coefficient U″(X)/U′(X)
$(1/\gamma)X(T)^{\gamma}$, $1 > \gamma \neq 0$	0	$-(1-\gamma)/X$
ln X(T)	$-\infty$	$-1/X$
$1-\exp(-\alpha X(T))$	0	$-\alpha$

The expected utility is (4.10). Divide by $U'(X) \neq 0$ and assume that $E(Y) = 0$. Therefore (4.11) follows.

$$E[U(X-Y)] = U(X) - U'(X)E(Y) + (1/2)(U''(X))E(Y^2) \tag{4.10}$$

$$\{E[U(X-Y)] - U(X)\}/U'(X) = (1/2)[U''(X)/U'(X)]E(Y^2) \tag{4.11}$$

The loss of utility due to the probability of losses depends upon risk aversion $U''(x)/U'(x)$, and risk σ^2. When $U''(X)/U'(X)$ is significantly negative, even if E(Y) is very small the variance imposes a large penalty.

$$\text{Risk Aversion} = -U''(X)/U'(X) = \text{the price one would pay for certainty.} \tag{4.12}$$

There are many criterion functions that satisfy these requirements. Only a few can be solved analytically in the optimization process. The others must be calculated numerically. In Table 4.1 I consider three widely used functions. The first is often called HARA. The second is a logarithmic function, which is a special case of HARA when $\gamma = 0$. The third is an exponential function.

These functions are different. In the HARA and exponential cases, values of X approaching zero are not severely penalized $U(0) = 0$. In the logarithmic case, low values of X(T) are very severely penalized. In the HARA and logarithmic cases, the risk aversion coefficient in column three depends negatively upon net worth. In the exponential case, it is independent of net worth.

There are several great advantages to using the logarithmic criterion function. First, in all cases one can derive the stochastic optimal control (SOC) – debt or liability ratio – by using dynamic programming. However, in the logarithmic case, the SOC can be derived directly by integrating the Ito equation. Second, a Mean-Variance interpretation can be given to the determination of the optimal control, as I show later in this chapter. Third, the risk coefficients either γ or α are arbitrary. Specifically, suppose that the debtor has a low risk aversion, $(1 - \gamma)$ is close to zero in the HARA case. His optimum debt ratio would be very high, close to risk neutrality. In the LTCM case, the market view was that LTCM "ran in front of bulldozers to pick up nickels". The lender is unlikely to be risk neutral. He may have a very high risk aversion, and be reluctant to accommodate the borrower's desired debt. So the choice of γ or α is arbitrary. There is a way to cope with this problem (Stein 2007) in a general equilibrium framework, where the rate of interest equilibrates the demand and supply of debt.

There is another criterion. It involves selecting a control that maximizes the expected minimum utility of net worth or consumption: max E[min U(X)]. This approach leads into a differential game against Nature (Fleming 2005a, b). This model is deterministic. It has a lot of merits, but the key parameters are arbitrary.

4.5 Methods of Solution of Stochastic Optimal Control Problem

The mathematics of solving for the optimal controls, which maximize the criterion function subject to the stochastic differential equations and constraints, are derived and discussed in Fleming–Rishel, Ch. V, VI, and Fleming and Stein (2004). In the appendix, I explain in detail the method of solution of the logarithmic case where dynamic programming is not needed. The Ito equation is sufficient. In this section, I relate the derivation, in Model II, to a Mean-Variance interpretation.

I use Model II as an example of SOC. Using (4.3) and (4.4), the stochastic differential equation for the change in net worth is (4.13).

$$\begin{aligned} dX(t) &= X(t)[A(f(t))\,dt + B(f(t))]dt \\ A(t) &= (1+f(t))b - if(t) - c \\ B(f(t)) &= (1+f(t))\sigma_b dw_b - f(t)\sigma_i dw_i \end{aligned} \tag{4.13}$$

The first term A(f(t)) is deterministic, call it the Mean. The second term B(f(t)) contains the two stochastic variables where the dw_i and dw_b terms are Brownian Motion (BM). The two BM terms are correlated. In the realistic case, the correlation is negative $E(dw_b dw_i) = \rho\, dt$, where $0 > \rho > -1$. A decline in the interest rate increases the demand for houses and raises the price. Similarly, a rise in the interest rate decreases the demand for houses and lowers the price.

The solution of (4.13) involves the stochastic calculus. The criterion is to maximize the expectation of the logarithm of net worth at a terminal date. The Ito equation for the change in logarithm of net worth d(ln X(t)) implies (4.14), the expectation of the change in net worth. Fleming and Rishel, and Øksendal are excellent sources that explain the derivation.

$$\begin{aligned} E[d(In\,X(t))] &= \{[M[f(t)] - R[f(t)]]\}dt \\ Mean\,M[f(t)] &= [(1+f(t))b - if(t) - c] \\ Risk\,R(f(t)) &= (1/2)[(1+f(t))^2\sigma_b^2 + f(t)^2\sigma_i^2 - 2f(t)(1+f(t))\sigma_b\sigma_i\rho] \end{aligned} \tag{4.14}$$

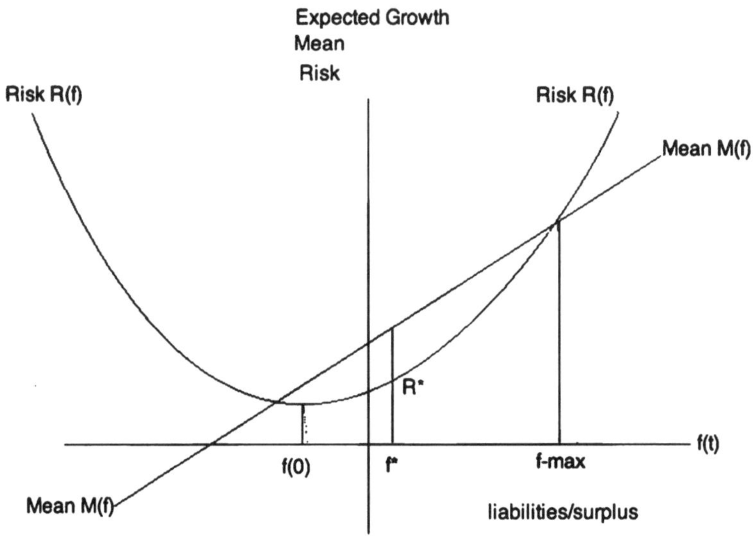

Fig. 4.1 Expected growth of net worth, E [d ln X(t)] = M[f(t)] − R[f(t)]. Variable f(t) = debt/ net worth or liabilities/surplus

The optimum debt/net worth f(t) maximizes the difference between the mean term M(f(t)) and the risk term R(f(t)). Figure 4.1 graphs the two terms against the debt ratio. The mean term is the straight line M(f) and the risk term R(f) is the quadratic term.

The mean line is the growth of net worth *if there were no risks*. The slope of this line is (b – i), the difference between mean return on assets and the mean interest rate. The intercept is (b − c), the mean return on assets less the ratio of consumption/net worth. Risk is a quadratic in the debt ratio. The slope depends upon σ_b, σ_i, ρ, the variances of the two stochastic variables and ρ the correlation between the two stochastic variables.

The optimum debt ratio is f*(t), where the difference between line M(f) and quadratic R(f) is maximal. It is

$$f^* = argmax_f[M(f(t) - R(f(t))] = [(b - i) - (\sigma_b^2 - \rho\sigma_i\sigma_b)]/\sigma^2 \tag{4.15}$$

$$\sigma^2 = \sigma_i^2 + \sigma_b^2 - 2\,\rho\,\sigma_i\,\sigma_b$$

It is positively related to (b−i) the mean return on capital less the mean interest rate and negatively related to the risk elements. The negative correlation between the two BM terms increases σ^2 the total risk.

The stochastic case in this book contrasts with the deterministic case. In the deterministic case there is no R(f) term. The economics literature assumes that the return on assets b(f) is a concave function. So more debt should be incurred as long as the slope of the term (b − i) is positive, that as long as M′(f) is positive.

The value of the debt ratio where (b(f) − i)f is maximal, where M′(f) = 0, may be very large. In the deterministic case, there is no concern that there will be unexpected "bad" realizations of (b, i) that will render the firm unable to service its debt – no matter how large.

4.6 Loss of Expected Growth from Misspecification

No one knows what is the correct model, or even the exact values of the parameters such as b and i. Suppose that the true model is described by Fig. 4.1 and the implied optimum debt ratio is f*(t). What happens if values (b_1, i_1) or (b_2, i_2) were used in the optimization (4.15), whereas the true unknown values are (b,i). Figure 4.2 describes the change in expected growth, which is the difference between the Mean M(f) and Risk R(f) in Fig. 4.1.

The expected growth of net worth E[d ln X(t)] is maximal at the optimal debt ratio f*(t). As the actual debt ratio rises above it, the expected growth declines. At debt ratio f-max, the expected growth is zero and above f-max it is negative. The actual growth is the expected value plus a random term.

One can never know what is the "true" model and hence what is the "true" optimal debt ratio f*(t). One chooses what seems to be the optimal debt ratio based upon what seems to be the correct model. There will always be a specification error at any time t measured as the excess debt $\Psi(t) = (f(t) - f^*(t))$.

Suppose that the true unknown model is described by the curve in Fig. 4.2. When f*(t) is selected, the expected optimum growth is maximal at W*(t). Given

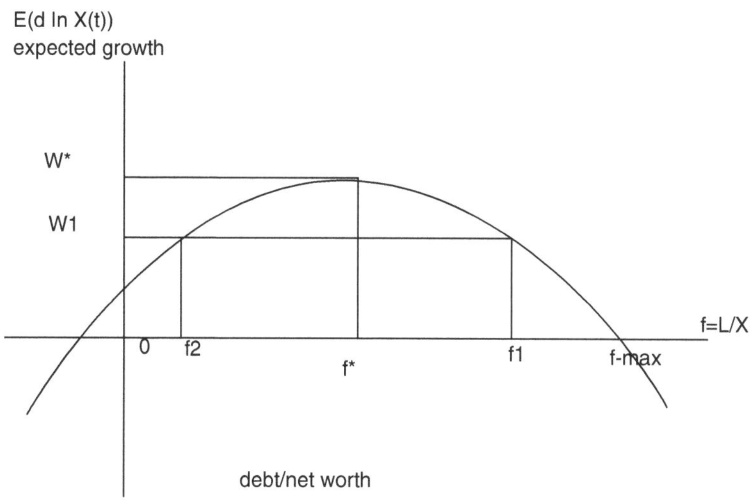

Fig. 4.2 The change in expected growth equal to the difference between the Mean M(f) and Risk R(f) in Fig. 4.1

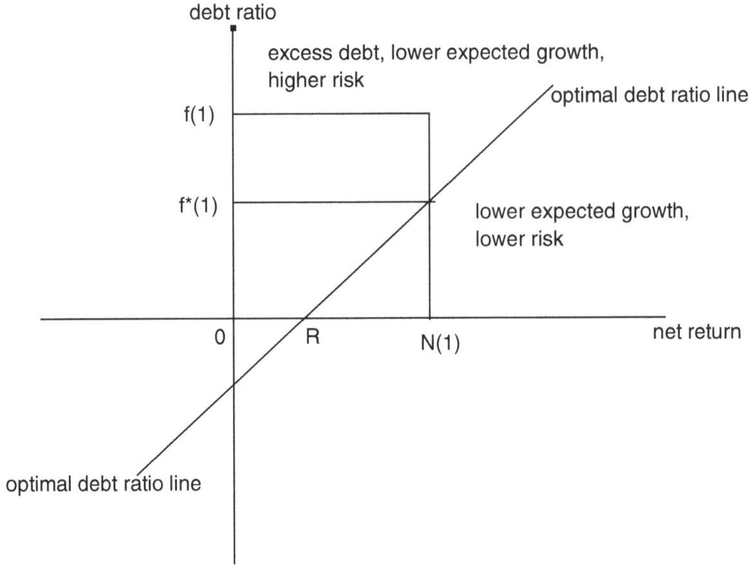

Fig. 4.3 Optimum debt ratio and loss of expected net worth from model mis-specification

limited information, the optimizing agents have faulty estimates of the key parameters (b, i). Some select f_1 and others select f_2 as their optimum debt ratios. Then, for example, the expected growth would be $W_1 = W_2 < W^*$. The loss of expected growth from model misspecification is $W^* - W_1$ in the cases drawn. Equation (4.16), which is the integral of (14),

$$E[lnX^*(T) - lnX(T)] = \int^T [W^*(t) - W(t)]dt = (1/2)\int^T \sigma^2 \Psi(t)^2 dt. \tag{4.16}$$

$$\sigma^2 = \sigma_i{}^2 + \sigma_b{}^2 - 2\rho\sigma_i\sigma_b$$

states precisely what is the loss of expected growth of net worth as a result of using a non-optimal control, debt ratio f(t). It is proportional to the square of the excess debt $\Psi(t)^2 = [f(t) - f^*(t)]^2$. Consequently, one can evaluate how sensitive the results, the expected growth of net worth, are to model specifications.

The relation between the optimal debt ratio and the fundamentals, (4.15), is described in Fig. 4.3. The debt ratio is on the ordinate. The fundamental, the mean net return (b − i), is on the abscissa. The slope of the line is $1/\sigma^2$.

If the net return were N(1), the optimal debt ratio is f*(1), which corresponds to f*(t) in Fig. 4.2. As the actual debt ratio rises above f*(1), the expected growth declines, as seen in Fig. 4.2. Risk, stated in (4.14), rises. If the debt ratio were less than f*(1), both the expected growth and risk decline.

The empirical work in the chapters below estimates the excess debt ratio, equal to the actual ratio less optimal debt ratio. Since there is model uncertainty, I derive an upper bound for the optimal ratio called f**(t), and calculate an excess debt as $\Psi(t) = [f(t) - f^{**}(t)]$. Then, on the basis of Fig. 4.2, the probability of a debt crisis is directly related to $\Psi(t)$.

The upper bound f**(t), discussed in Chap. 5, assumes in Model I that the mean interest rate is equal to the trend rate of growth of the price and there is no deviation of the price from trend. In Model II the upper bound f**(t) assumes that $(b - i) = \beta(t)$ the productivity of capital.

4.7 Insurance

4.7.1 Cramér-Lundberg

In the AIG case, the question is how to estimate the optimal amount of insurance to offer. I follow Volchan's (2007) exposition and use his notation. He writes that the Cramér/Lundberg (C-L) model is still a landmark of non-life insurance mathematics. It is to actuarial science what the Black-Scholes-Merton model is to finance. The decision process concerns the amount of reserve capital that the insurance company should keep in order to absorb the impact of possible losses due to adverse changes in value of claims. The basic equation is (4.17), used in all variants.

$$\begin{aligned} \textit{Change in surplus} = {} & (\textit{premium rate})\textit{liabilities} + (\textit{return} + \textit{capital gain})\textit{assets} \\ & - (\textit{claims on the company related to liabilities}) \end{aligned} \tag{4.17}$$

The C-L model and its variants are concerned with the ruin problem, to be discussed below. In (4.18), U(t) is the surplus, equal to initial capital u, plus total premium income $\Pi(t)$ less aggregate claims S(t). The deterministic total of premium income $\Pi(t)$ in (4.19) is the product of premium rate $c > 0$ and time t. The aggregate claims S(t) is the sum of $X_k(t)$, amounts to be paid out in the period (4.20). The number of claims arriving in the interval (0,t) is N(t).

$$U(t) = u + \Pi(t) - S(t) \tag{4.18}$$

$$\Pi(t) = ct. \tag{4.19}$$

$$S(t) = \sum X_k(t), k = 0, 1, \ldots N(t). \tag{4.20}$$

The crucial assumption of the C-L model is that the claim number process {N(t)} is a homogeneous Poisson process equation (4.21). This implies that the probability that k claims will arrive in the interval is Poisson distributed. Thus $P(0, \lambda t) = e^{-\lambda t}$ is the probability of no claims arriving during the interval of length t. The aggregate claims {S(t)} is a compound Poisson process described by

$$P\{(N(t) = k\} = e^{-\lambda t}(\lambda t)^k/k! \tag{4.21}$$

The focus is upon (4.22) the probability Θ(t) of bankruptcy/insolvency, that the surplus at time t will be non-positive.

$$\Theta(t) = P[U(t) - u < 0], \textit{for some } t > 0. \tag{4.22}$$

The solution has the following characteristics. Using the above equations the expected value of the surplus [EU(t) − u] is (4.23). The expected number of claims E[N(t)] = λt and μ is the average cost per claim. Total premiums are ct. The *net profit condition* is that (c − λμ) > 0, the premium must exceed the average total claim.

$$E[U(t)] - u = ct - \mu E[N(t)] = t(c - \lambda\mu). \tag{4.23}$$

Even though the premium c > λμ, the process of U(t) can attain negative values in some subperiods. So it is important to have at least an estimate of Θ(t) = P[U(t) − u] < 0 guaranteeing that it is sufficiently small.

The Cramér-Lundberg analysis is inadequate to evaluate the risk/return in the AIG case. First, the asset side of the equation for the change in surplus is ignored. The insurance company has assets against the liabilities that bring in income. This is the second term in (4.17). Second, the value of the claims against AIG are highly negatively correlated with the value of the assets. When the market value of the insured assets decline, AIG must either compensate the insurer for the difference or put up more collateral. Third, the assets that are insured by AIG are quite closely correlated with the assets in AIG's portfolio.

4.7.2 Ruin Analysis

The C-L model is closely related to the Ruin problems in the probability literature. See Feller I, II. The ruin literature is concerned with the question, what is the probability that a firm that starts with a capital z will ultimately be ruined? Thereby one has a framework to determine "optimal" capital.

The structure of the Ruin model is simple. In each period, a firm with a capital of z can either gain $k with probability 1 > p > 0, or lose k with probability q = 1−p. Say that the firm has borrowed $a. If the capital falls to zero, the firm is bankrupt: the probability of ruin is then one. If the capital rises to $a, the probability of ruin is zero. The firm repays its debt and the story can start over.

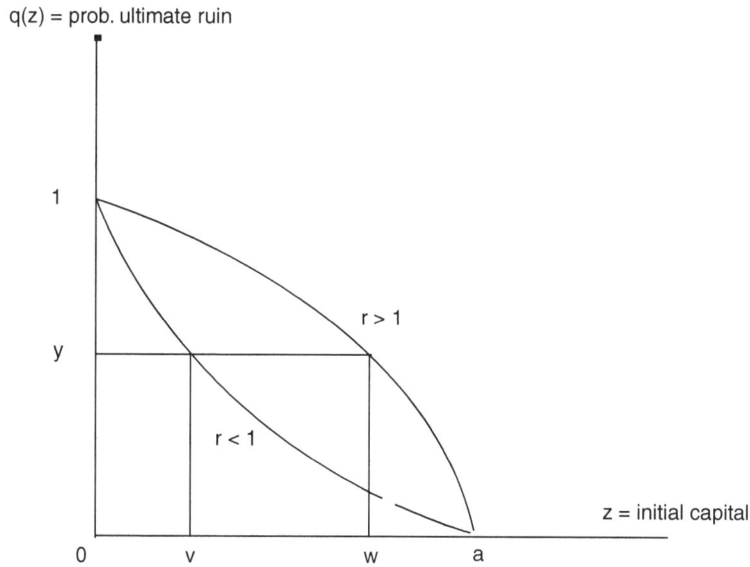

Fig. 4.4 Probability of ultimate ruin

Let Q(z) be the probability that a firm with a initial capital of z will go bankrupt, before repaying its debt. The terminal conditions are: Q(0) = 1, Q(a) = 0. The solution of the stochastic difference equation for Q(z) is (4.24), where r = q/p is key parameter.

$$Q(z) = \left[r^{a/k} - r^{z/k}\right] / \left[r^{a/k} - 1\right], \qquad r = q/p \neq 1 \tag{4.24}$$

Figure 4.4 graphs this equation for two values r > 1, and r < 1. At z = 0, Q(0) =1 and at z = a, Q(a) = 0. The probability of ultimate ruin is clearly greater for r > 1 than for r < 1. If the "target" probability of ultimate ruin is y, then "optimal" initial capital is z = v when r < 1, and z = w when r > 1. The greater the odds of a gain relative to a loss, the smaller is the required initial capital.

4.7.3 *The Stochastic Optimal Control Approach to Insurance*

The SOC analysis explained above is applied to the AIG case. Unlike the ruin problem or the Cramér-Lundberg approach, the criterion does not focus solely upon the probability of ruin. The criterion function, the maximizaton of the expected logarithm of surplus X(T) at a terminal date T, contains both the growth of net worth and risk aversion. The logarithm function implies that an infinite penalty is placed upon bankruptcy. In this manner the SOC analysis contains the best points of the other approaches.

The dynamics of the process to derive E[ln X(T)] is based upon (4.25). There are several possibilities for modeling the stochastic processes for the claims C(t) and capital gains dP(t)/P(t). Assets = A(t). Insurance liabilities = L(t).

$$dX(t) = [\pi L(t)dt + \beta A(t)dt] + [(dP(t)/P(t))A(t)] - C(t). \tag{4.25}$$

Assume, for simplicity, that the premium rate π is given. At this premium rate there is an elastic demand for insurance and the insurer decides how much insurance L(t) to offer at that rate. Income from premiums is π *L(t)dt*. The insurer has A(t) of assets whose value is very closely related to the assets that it has insured, the CDOs in particular. The return on these assets has two parts, a rate of return β that is deterministic and a future capital gain or loss *(dP(t)/P(t))A(t)* that is unknown. The most important stochastic variable is C(t), the future claims against the insurer against the liabilities – insurance policies.

The modeling consists of specifying the stochastic processes on the capital gain and the claims. Let the claims C(t) be described by stochastic differential equation (4.26). They are proportional to L(t) the amount of insurance liabilities. Claims are the required payments to the insured holders of CDS, due to either defaults of the obligors or for collateral calls when the prices of the insured securities decline. The latter led to the downfall of AIG. The mean of claims C(t) is cL(t)dt. The variance of the claims is $\sigma_c^2 L^2(t)$ dt. Brownian Motion term dw_c has independent and stationary increments, with zero expectations, as assumed in the classical literature.

$$\begin{aligned} C(t) &= [cdt + \sigma_c dw_c]L(t). \\ E(dw_c) &= 0, E(dw_c)^2 = dt. \end{aligned} \tag{4.26}$$

Stochastic differential equation (4.27) concerns the capital gain/loss term dP(t)/P(t). The time varying drift is a(t)dt and the diffusion is $\sigma_p dw_p$. This is a general formulation. Models 1 and 2 of the asset price or capital gain, discussed above, imply different values of the drift a(t). The variance of the capital gain is σ_p^2 dt. The dw_p term has independent and stationary increments.

$$\begin{aligned} dP(t)/P(t) &= a(t)dt + \sigma_p dw_p \\ E(dw_p) &= 0,\ E(dw_p)^2 = dt. \end{aligned} \tag{4.27}$$

The two Brownian motion terms are expected to be negatively correlated (4.28): correlation coefficient $-1 \leq \rho < 0$. When there is a period with capital losses dP(t)/P(t) < 0, then claims against AIG including collateral calls are most likely to be high. This condition accurately described the period of the AIG crisis with CDS in Chap. 6.

$$E(dw_p dw_c) = \rho\, dt. \quad -1 \leq \rho < 0 \tag{4.28}$$

Using (4.26), (4.27), and (4.28) in (4.25), derive the stochastic differential equation (4.29) for the change in surplus dX(t). The first set of brackets contains the deterministic part and the second *set contains the stochastic part.*

$$dX(t) = [\pi L(t) + \beta A(t) - cL(t) + A(t)a(t)]dt + [A(t)\sigma_p dw_p - L(t)\sigma_c dw_c]. \quad (4.29)$$

X = A-L = surplus, A = assets, L = insurance liabilties, β = return on assets, π = premium rate, C = claims, a(t) = drift capital gain, stochastic BM terms dw_p, dw_c. Correlation between them is $\rho < 0$. Leverage = A/X = (1 + f), debt ratio f = L/X.

The SOC analysis follows very closely to what was done in Sects. 4.5 and 4.6 above. The optimal ratio of insurance liabilties/surplus f*(t) is (4.30).

$$f^*(t) = [(\pi - c) + (\beta + a(t))] - \left(\sigma^2{}_p - \rho\sigma_p\sigma_c\right) / \left(\sigma^2{}_p + \sigma^2{}_c - 2\rho\sigma_p\sigma_c\right) \quad (4.30)$$

The first term in brackets in (4.30) for the optimum ratio is (π − c) the premium less the drift of the claims plus (β + a(t)) the sum of the rate of return on assets plus the time varying drift of the capital gain. The second term and the denominator involve the risks of the capital gains and claims. The negative correlation between the capital gain and the claims increases total risk and reduces the optimal ratio of insurance liabilities/surplus.

Figures 4.1, 4.2, and 4.3 describing the determination of the loss of growth from model misspecification and the relation between the net return and the optimal ratio apply equally here for the optimal insurance/surplus.

AIGFP relied on Gorton's actuarial model that did not provide a tool for monitoring the CDO's market value. Gorton's model (FCIC, p. 267) had determined with 99.85% confidence that the owners of super-senior tranches of the CDO's insured by AIGFP would never suffer real economic losses. The company's auditor PWC apparently was also not aware of the collateral requirements. PWC concluded that "..the risk of default on the [AIG] portfolio has been effectively removed as a result of a risk management perspective...".

The SOC analysis leads to a very different way of viewing AIG risk than was used by Gorton and the company's auditor PWC.

4.8 The Endogenous Changing Distributions

Underlying all dynamic optimizations is the characterization of the price distributions. The usual assumptions are that the system parameters are unchanging. Ren Cheng (Fidelity Investments) gave an insightful presentation of the limitations of the existing approaches used in financial modeling. I conclude this chapter with a discussion of Cheng's presentation.

The finance models assume that the means-variances-covariances are stable. Each entity in the optimization is considered in isolation. There is no consideration of a feedback among units, or the global effects of the collective action upon the distribution functions. The latter are assumed to be normal, an assumption justified by the Central Limit Theorem (CLT). We know a lot about the decision making process of an individual unit in portfolio optimization. But we are ignorant of how the units interact and thereby change the distributions. In Chaps. 5 and 6, I explain how the interactions among units in the shadow banking system magnified the collapse of housing prices upon the financial system. The whole was much greater than the sum of its parts.

The most critical requirement for the CLT, and for almost all of the distributions used in probability theory, is the independence. Normality of sample distributions follows from the CLT. With dependence and feedbacks, the tail gets longer and heavier. The nature of financial markets changes as a result of the collective actions.

I can formalize Cheng's views by showing how the ruin theorems and Cramér-Lundberg model are changed by interactions. In the ruin model Sect. 4.7.2, the object is to determine the probability Q(z) of being ruined as a function of capital z. When $z = 0$ the probability $Q(0) = 1$. When $z = a > 0$, then $Q(a) = 0$. The firm repays the debt and the process restarts. At each time the probability of gaining \$1 is $1 > p > 0$ and $q = 1 - p$ is the probability of losing \$1. The basic equation for the change in probability is

$$Q(z) = Q(z-1)p + Q(z+1)q \tag{4.31}$$

Thereby an equation like (4.24) was derived in the Gambler's Ruin problem.

Consider an alternative to either this model or to the Bernoulli process that underlies the Binomial distribution. In the Bernoulli process there is an idealized urn with *b* black and *r* red balls. Sampling is done with replacement. The probability of drawing a black ball at any time is constant at $p = b/(b + r)$. Insofar as the samples are independent, the Bernoulli case, the probability of drawing k black balls in $n > k$ trials is $Pr(k;n,p) = C(n,k)\, p^k q^{n-k}$, where C(n,k) is the factorial term.

An alternative is an *Urn process*. When a ball is drawn, then c balls of the same color are added. Calculate the probability that at the nth trial there will be k black balls. The contrast between the two approaches is easily seen in an example. What is the probability of drawing two black balls in two trials Pr(BB)? In the *Bernoulli independence* case, since $Pr(B) = p = b/(b + r)$. It follows that

$$Pr(BB) = Pr(B)Pr(B) = p^2. \tag{4.32}$$

In the *Urn process* the probability of drawing two black balls is $Pr(BB) = Pr(B|B)\, Pr(B)$. The probability of drawing a black ball on the first trial is $Pr(B) = b/(b + r)$. The urn now contains (b + c) black balls and r red balls. The conditional probability of drawing a black ball on the second trail is $Pr(B|B) = (b + c)/(b + r + c) > b/(b + r)$. Therefore the conditional probability is greater than the marginal probability. As a result

$$Pr(BB) = [(b+c)/(b+r+c)][b/(b+r)]. \tag{4.33}$$

Think of a black ball as a success, and a red ball is a failure. Then the Urn model implies that: Success breeds another success and a failure breeds another failure. The conditional probability is greater than the marginal probability.

In the financial market, if prices rise then firms buy and stimulate further price rises. A black ball drawn leads to c black balls added. If a red ball is drawn this indicates a decline in price. When prices fall firms sell which stimulates further declines. One adds c red balls. Conditional probabilities change endogenously. The history of the financial crisis in Chaps. 5 and 6 is consistent with this alternative model. Actions by firms change the probability distributions.

I conclude by repeating Derman's view. "Models are only models, not the thing in itself. We cannot therefore expect them to be truly right. Models are better regarded as a collection of parallel thought universes you can explore. Each universe should be consistent, but the actual financial and human world, unlike the world of matter, is going to be infinitely more complex than any model we make of it. We are always trying to shoehorn the real world into one of the models to see how useful approximation it is".

4.9 Mathematical Appendix

I show how the optimal debt ratio is derived in the logarithmic case.

The stochastic differential equation is (A1).

$$dX(t) = X(t)[(1 + f(t))(dP(t)/P(t) + \beta(t)dt) - i(t)f(t) - c\ dt] \qquad \text{(A1)}$$

X(t) = Net worth, f(t) = debt/net worth = L(t)/X(t), dP(t)/P(t) = capital gain or loss, i(t) = interest rate, (1 + f(t)) = assets/net worth, β(t) = productivity of capital. C(t)/X(t) = c (t) = consumption/net worth = c is taken as given.

Let the price evolve as

$$dP(t) = P(t)\big(a(t)\,dt + \sigma_p dw_p\big) \qquad \text{(A2)}$$

where drift a(t) will depend upon the Model I or II. The interest rate evolves as

$$i(t) = idt + \sigma_i dw_i \qquad \text{(A3)}$$

Substitute (A2) and (A3) in (A1) and derive (A4)

$$dX(t) = X(t)\{[(1 + f(t))(a(t)dt + \beta(t)dt) - (\mathrm{if(t)dt + c\ dt})]$$

$$+ \big[(1 + \mathrm{f(t)})\ \sigma_\mathrm{p}\mathrm{dw_p} - \ \sigma_\mathrm{i}\mathrm{f(t)dw_i}\big]\}$$

$$dX(t) = X(t)M(f(t)dt + X(t)B(f(t) \qquad \text{(A4)}$$

$$M(f(t)) = [(1 + f(t)(a(t)dt + \beta(t))dt - (if(t) + c)]$$

$$\mathrm{B(t) = [(1 + f(t))\sigma_p dw_p - \sigma_i f(t) dw_i]\}}$$

$$B^2(f(t) = (1 + f(t)^2 \sigma^2{}_p dt + f(t)^2 \sigma^2{}_i dt - 2f(t)(1 + f(t))\sigma_i \sigma_p dw_p dw_i$$

$$Risk = R(f(t)) = (1/2)\left[(1 + f(t))^2 \sigma_b{}^2 + f(t)^2 \sigma_i{}^2 - 2f(t)(1 + f(t))\sigma_b \sigma_i \rho)\right]$$

M(f(t)) contains the deterministic terms and B(f(t)) contains the stochastic terms. To solve for X(t) consider the change in ln X(t) (A5). This is based upon the Ito equation of the stochastic calculus. A great virtue of using the logarithmic criterion is that one does not need to use dynamic programming. The expectation of d ln X(t) is *(*A6*)*.

$$dlnX(t) = (1/X(t))dX(t) - \left(1/2X(t)^2\right)(dX(t))^2 \tag{A5}$$

$$E[d(lnX(t))] = [M[f(t)] - R[(f(t)]dt] \tag{A6}$$

The correlation ρ dt $=$ E($dw_p dw_i$) is negative, which increase risk. $(dt)^2 = 0$, dwdt $= 0$.

The optimal debt ratio f* maximizes the difference between the Mean and Risk curves in Fig. 4.1.

$$f* = argmax_f\ [M(f(t)) - R(f(t))] = [a(t) + \beta(t) - i] - [(\sigma_p{}^2 - \rho\sigma_i\sigma_b)]/\sigma^2 \tag{A7}$$

$$\sigma^2 = \sigma_i{}^2 + \sigma_p{}^2 - 2\,\rho\,\sigma_i\,\sigma_p$$

Model I

Model I assumes that the price P(t) has a trend rt and a deviation y(t) from it (A8). The deviation y(t) follows an Ornstein-Uhlenbeck ergodic mean reverting process (A9). Coefficient α is positive and finite. The interest rate is the same as in Model II.

$$P(t) = P(0)exp[rt + y(t)]. \tag{A8}$$

$$dy(t) = -\alpha y(t) + \sigma_p dw_p \tag{A9}$$

Therefore using the stochastic calculus a(t) in Model I is the first term in (A10),

$$dP(t)/P(t) = (r - \alpha y(t) + (1/2)\sigma_p{}^2)dt + \sigma_p dw_p. \tag{A10}$$

Substitute (A10) in (A7) and derive (A11), the optimal debt ratio in Model I.

$$f^*(t) = [(r - i) + \beta - \alpha y(t) - (1/2)\sigma_p{}^2 + \rho\sigma_i\sigma_p]/\sigma^2. \qquad \text{(A11)}$$

Consider $\beta(t)$ as deterministic.

Model II

In Model II, the price equation is (A12). The drift is a(t) dt $= \pi$ dt and the diffusion is $\sigma_p dw_p$.

$$dP(t)/P(t) = \pi dt + \sigma_p dw_p. \qquad \text{(A12)}$$

The optimal debt ratio f*(t) is (A13). Consider $\beta(t)$ as deterministic.

$$f^*(t) = [(\ \pi + \ \beta(\mathrm{t}) - \mathrm{i}) - (\ \sigma^2{}_\mathrm{p} - \ \rho\ \sigma_\mathrm{i}\ \sigma_\mathrm{p})]/\ \sigma^2. \qquad \text{(A13)}$$
$$\sigma^2 = \ \sigma_\mathrm{i}{}^2 + \ \sigma_\mathrm{p}{}^2 - 2\ \rho\sigma_\mathrm{i}\sigma_\mathrm{p}$$

References

Cheng, Ren. 2011. Fat tail genesis, power point presentation, fidelity asset allocation.

Derman, E. 2004. *My life as a quant*. Hoboken: Wiley.

Feller, William. (I, 1950) (II, 1966) *An introduction to probability theory and its applications*. New York: Wiley.

Fleming, Wendell H. 2005a. Optimal investment models with minimum consumption criteria. *Australian Economic Papers* 44: 307–321. Special Issue on Stochastic Models.

Fleming, Wendell H. 2005b. Optimal investment models with minimum consumption criteria. *Australian Economic Papers* 44(4): 307.

Fleming, Wendell H., and R. Rishel. 1975. *Deterministic and stochastic optimal control*. Berlin: Springer.

Fleming, Wendell H., and M. Soner. 2006. *Controlled Markov processes and viscosity solutions*. New York: Springer.

Fleming, Wendell H., and Jerome L. Stein. 2004. Stochastic optimal control, international finance and debt. *Journal of Banking and Finance* 28: 979–996.

Øksendal, B. 1995. *Stochastic differential equations*. Berlin: Springer.

Stein, Jerome. 2004. Stocastic optimal control modeling of debt crises. In *Mathematics of finance, contemporary mathematics*, ed. G. Yin and Q. Zhang. Providence: American Mathematical Society.

Stein, Jerome. 2005. Optimal debt and endogenous growth in models of international finance. *Australian Economic Papers* 44(4): 389.

Stein, Jerome. 2007. United States current account deficits: A stochastic optimal control approach. *Journal of Banking and Finance* 31(5): 1321.

Volchan, S. 2007. The fundamental theorem of actuarial risk science. *The Mathematical Scientist* 31(2): 73–84.

Chapter 5
Application of Stochastic Optimal Control (SOC) to the US Financial Debt Crisis

Abstract The financial crisis was precipitated by the mortgage crisis. A whole structure of financial derivatives was based upon the ultimate debtors, the mortgagors. Insofar as the mortgagors were unable to service their debts, the values of the derivatives fell. The financial intermediaries whose assets and liabilities were based upon the value of derivatives were very highly leveraged. Changes in the values of their net worth were large multiples of changes in asset values. A cascade was precipitated by the mortgage defaults. In this manner, the mortgage debt crisis turned into a financial crisis. The crucial variable is the optimal debt of the real estate sector, which depends upon the capital gain and the interest rate. I apply the SOC analysis to derive the optimal debt. Two models of the stochastic process on the capital gain and interest rate are presented. Each implies a different value of the optimal debt/net worth. I derive an upper bound of the optimal debt ratio, based upon the alternative models. An empirical measure of the excess debt: actual less the upper bound of the optimal ratio, is shown to be an early warning signal (EWS) of the debt crisis. Finally, the shadow banking system is discussed.

5.1 Introduction

The mortgage/housing market had increased in importance in financial markets from 1985 to 2005. The assets of the banking sector and financial intermediaries were closely and significantly tied to the real estate market. Contrary to the Efficient Market Hypothesis (EMH), there was indeed a mortgage market bubble, and prices did not reflect publicly available information on fundamentals. I measure the "bubble" in terms an unsustainable debt/income ratio of households. Quants and the market missed systemic risk and ignored publicly available information concerning the debt ratios of the households. Using Stochastic Optimal Control (SOC) analysis in Chap. 4, I derive an optimal debt ratio. An Early Warning Signal (EWS) of a debt

J.L. Stein, *Stochastic Optimal Control and the U.S. Financial Debt Crisis*,
DOI 10.1007/978-1-4614-3079-7_5, © Springer Science+Business Media New York 2012

crisis is the excess debt, defined as the difference between the actual and optimal debt ratio of households. The excess debt determines the probability of a debt crisis.

When the households were experiencing difficulties in servicing their debts, the highly leveraged banks and financial intermediaries suffered significant losses and their net worth declined to an alarming degree. This was the financial crisis of 2008. In the S&L and agricultural crises of the 1980s, discussed in Chap. 7, the mortgages were not a very important part of portfolios of banks and financial intermediaries. Hence these earlier crises did not have significant effects upon the US financial system.

Section 5.1 concerns the importance of the housing sector to the financial sector. Section 5.2 concerns the characteristics of the mortgage market. The SOC analysis concerning the optimal debt ratio is in Sect. 5.3. An economic interpretation of this is the subject of Sect. 5.4. Section 5.5 concerns empirical measures of the optimal debt ratio. Early Warning Signals (EWS) of a crisis are presented in Sect. 5.6. The model used by the market, which led to the crisis, is the subject of Sect. 5.7. Section 5.8 focuses upon the Shadow Banking System, leverage and financial linkages.

5.2 The Importance of the Housing/Mortgage Sector to the Financial Sector

Juselius and Kim show that financial obligations ratios, rather than leverage, can explain financial fragility. They examine different loan categories of banks and infer that the compositional dynamics in credit losses of banks, resulting from the business and household real estate sectors, may be of particular importance for understanding financial fragility. The recent crisis was predominantly caused by too large financial obligations/mortages of the household sector to the banks and intermediaries, whereas the recession in the 1980s was more related to the business sector.

Table 5.1 describes the changing composition of bank portfolios. Real estate loans rose from 30% in 1985q1 to 57% in 2005q1. Moreover, the share of households in the banks' real estate loans rose from 48% in 1985q1 to 58% in 2005q1.

Table 5.1 Banks's aggregate portfolio

	1985q1	1995q1	2001q1	2005q1
Share total loans				
1. Real estate	30%	46%	47%	57%
2. Business loans	34	23	26	20
3. Other	36	31	27	23
Share R.E. loans				
4. Households	48	59	56	58
5. Business	26	29	28	26
6. Other	26	12	16	16

Source: Juselius and Kim (2011)

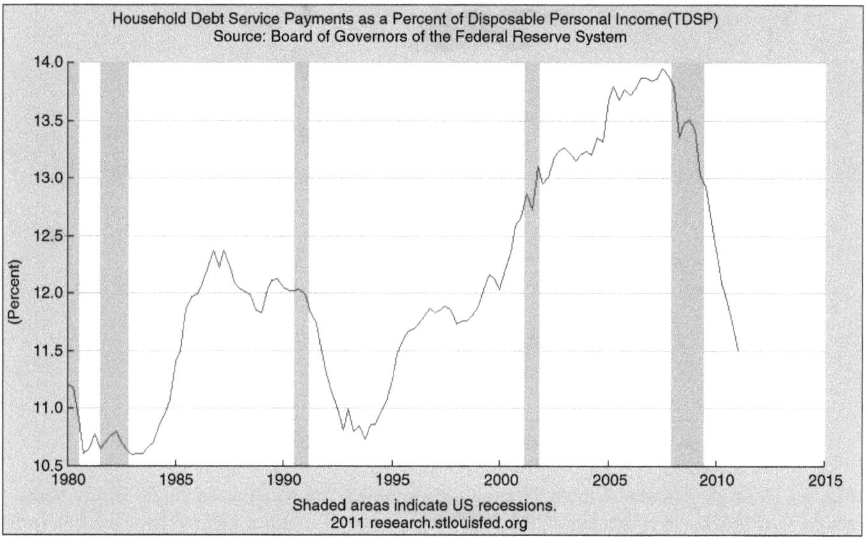

Fig. 5.1 Household debt service payments as a percent of disposable personal income, 1980–2011 (*Source*: Federal Reserve of St. Louis, FRED, from Federal Reserve)

An important finding is that the time path of the loss rate on real estate loans resembles that of the bank failure rate to a much greater extent than does the loss rate on business loans. They infer that financial stability was more sensitive to real estate loans than to business loans in the recent period. They write: "As is apparent, there are only two episodes of major bank failures in our sample. The first occurs between the late 1980s and early 1990s, and is associated with the savings and loan crisis, whereas the second corresponds to the recent financial crisis. It is notable that the bank failure rate is very low between these two periods, suggesting that the burst of the dot.com bubble had little effect on the incidence of bank failures".

5.3 Characteristics of the Mortgage Market

Demyanyk and Van Hemert (2008) utilized a data-base containing information on about one half of all subprime mortgages originated between 2001 and 2006. They explored to what extent the probability of delinquency/default can be attributed to different loan and borrower characteristics and housing price appreciation. Figure 5.1 plots Household debt service/disposable income. From 1998 to 2005 rising home prices produced above average capital gains, which increased owner equity. This induced a supply of mortgages, and the totality of household financial obligations as a percent of disposable personal income rose. The rises in house prices and owner equity induced a demand for mortgages by banks and funds.

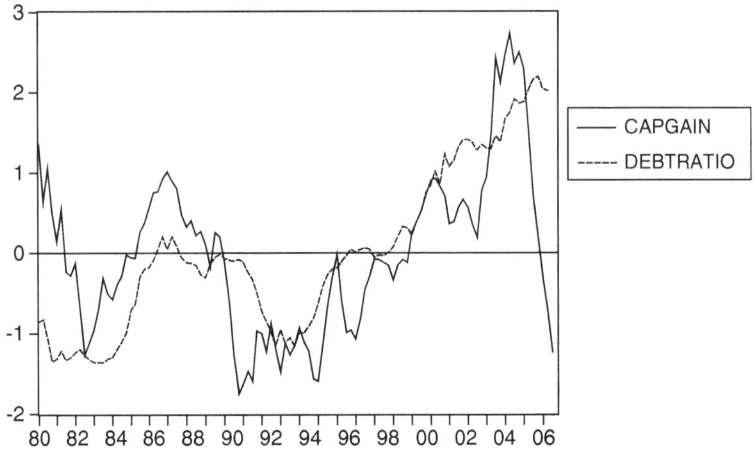

Fig. 5.2 Mortgage market bubble. Normalized variables. Appreciation of single-family housing prices, CAPGAIN, 4q appreciation of US Housing prices HPI, Office Federal Housing Enterprise Oversight (OHEO); Household debt ratio DEBTRATIO = household financial obligations as a percent of disposable income (Federal Reserve Bank of St. Louis, FRED, Series FODSP. Sample 1980q1–2007q4)

In about 45–55% of the cases, the purpose of the subprime mortgage taken out in 2006 was to extract cash by refinancing an existing mortgage loan into a larger mortgage loan. The quality of loans declined. The share of loans with full documentation substantially decreased from 69% in 2001 to 45% in 2006. Funds held packages of mortgage-backed securities either directly as asset-backed securities or indirectly through investment in central funds. The purchases were financed by short-term bank loans. Neither the funds nor the banks worried about the rising debt, because equity was rising due to the rise in home prices.

Figure 5.2 plots the capital gain and debt ratio. They are normalized so that each variable has a mean of zero and standard deviation of unity. Thus the vertical axis plots a t-statistic. The figure shows that from 2003 to 2005 the capital gain was more than two standard deviations above its long term mean, and the debt ratio was about two standard deviations above its long term mean. Since the mean interest rate over the entire period was about equal to the mean capital gain, a value of $t = 2$ should have been a warning signal that the capital gain cannot continue to exceed the mean interest rate. The "free lunch", described in Chap. 3, could not continue. Once the capital gain fell below the interest rate, the foreclosures and delinquencies would have led to the collapse of the value of the financial derivatives. The Quants ignored this, as explained in Chap. 3.

This view is confirmed in the study by Demyanyk and Van Hemert who had a data base consisting of one half of the US subprime mortgages originated during the

Table 5.2 Contribution C(i) of factors to probability of delinquency and defaults 2006, relative to mean for the period 2001–2006

Variable X(i)	Contribution C(i) to delinquency rate	Contribution C(i) to foreclosure rate
House price appreciation	1.08% = 7.2	0.61% = 15.250
Balloon	0.18 = 1.2	0.09 = 2.25
Documentation	0.16 = 1.07	0.07 = 1.75
Debt/income	0.15 = 1	0.04 = 1

Source: Demyanyk and Van Hemert (D-VH) 2007, Tables 2 and 3 for definitions of variables

period 2001–2006. At every mortgage age, loans originating in 2006 had a higher delinquency rate than in all the other years since 2001. They examined the relation between the probability Π of delinquency/foreclosure/binary variable z, denoted as $\Pi = \Pr(z)$ and sensible economic variables, vector X. They investigated to what extent a logit regression $\Pi = \Pr(z) = \Phi(\beta X)$ can explain the high level of delinquencies of vintage 2006 mortgage loans.

A logit model specifies that the probability that z = 1 is: $\Pr(z = 1) = \exp(X\beta)/[1 + \exp(X\beta)]$. Hence $\ln \{\Pr(z = 1)/\Pr(z = 0)\} = X\beta$. Vector β is the estimated regression coefficients. They estimated vector β based upon a random sample of one million first-lien subprime mortgage loans originated between 2001 and 2006. The first part of their study provides estimates of β, the vector of regression coefficients telling us the importance of the variables in vector X. The second part inquires why the year 2006 was so bad. The approach is based upon (5.1). The contribution C(i) of component X_i in vector X to why the probability of default in year 2006 was worse than the mean is:

$$C(i) = (\delta\Pi/\delta X_i)\, dX_i = \Phi(\beta X_m + \beta_i\, dX_i) - \Phi(\beta X_m) \tag{5.1}$$

X_m = mean value

The probability of delinquency/foreclosure when the vector X is at its mean value is $\Phi(\beta X_m)$. The added probability resulting from the change in component X_i in 2006 comes from $\beta_i dX_i$ where β_i is the regression coefficient of element X_i whose change was dX_i.

Table 5.2 (based upon D-VH, Table 3) displays the largest factors that made the delinquencies and foreclosures in year 2006 worse than the mean over the entire period. For year 2006, the largest contribution to delinquency and to foreclosure was the low house price appreciation. The absolute contributions are small, but the relative contributions are significant. The house price appreciation factor was 7 times the contributions of the debt/income and documentation ratios for the delinquency rate, and 15 times the contributions of the debt/income and documentation ratios for the foreclosure rate. The work of D-VH confirms the analysis above.

5.4 The Stochastic Optimal Control Analysis

As explained in the earlier chapters, the financial structure rested upon the ability of the mortgagors to service their debts. That is where "systemic risk" was to be found. When the households could not service their debts, the highly leveraged financial structure collapsed. For this reason the focus is upon the optimal debt ratio of households.

The stochastic optimal control approach was discussed and the basic propositions were derived in Chap. 4. In the present chapter I draw upon the results of that chapter. The hypothetical/idealized optimizer is the investor in real estate who finances his purchase with a mortgage debt. The optimum debt ratio is selected to maximize the expectation of a concave function of net worth at some terminal date, subject to stochastic processes on the house price and interest rate. The optimum debt/net worth ratio depends most crucially upon the assumed stochastic process concerning the price or capital gains variable. I explained in Chap. 4 that the expected growth of net worth is maximal when the debt ratio is optimal. As the debt ratio exceeds the optimal, the expected growth declines and the variance of growth increases. The excess debt is the actual less the optimal. As the excess debt rises, the probability that the household cannot pay its debt increases. Thereby we have an early warning signal.

Since no one knows what is the "true" stochastic process there is uncertainty concerning the true optimal debt ratio. In Chap. 4, I showed that the difference between the actual debt ratio and the true optimal is directly related to the difference between the actual growth of net worth and the optimal growth. This way one can have bounds for the effect of misspecification of stochastic process upon growth of net worth.

Several stochastic processes in Chap. 4 were considered and the optimal debt ratio was derived for each case. In the present chapter I simply state the results: the optimal debt ratio that is implied by the alternative stochastic processes, Model I and Model II. On the basis of Models I/II, I derive an equation for an empirical estimate of the optimal debt ratio. That is the benchmark optimal ratio. I then derive empirical estimates of the excess debt and show that it is a warning signal of a crisis. In this manner, I reject the Greenspan et al. view that the crisis was unpredictable.

Net worth X(t) is assets A(t) less liabilities L(t). Assets equals net worth plus debt. Equation 5.2 or 5.2a is the stochastic differential equation for the growth of net worth, the change in assets less the change in liabilities. The change in net worth is the sum of several terms in (5.2a). The first term is assets times its return. The return consists of the capital gain plus the productivity of the assets. The second term is the debt payments, the interest rate i(t) times L(t) the debt. The last term is consumption or dividends. Let f(t) = L(t)/X(t) denote the ratio of debt/net worth. The capital gain is dP(t)/P(t) where P(t) in our case represents an index of house prices. The productivity of capital $\beta(t)$ is income/assets, which is deterministic or constant; and c is the constant ratio of consumption or dividends/ net worth. The change in net worth is stochastic differential equation (5.2). Capital gain and interest rate are stochastic in the models.

$$\text{Change in net worth} = \text{assets (capital gain + productivity of assets)} \\ - \text{interest rate (debt)} - \text{consumption.} \qquad (5.2a)$$

$$dX(t) = X(t)[(1 + f(t))\,(dP(t)/P(t) + \beta(t)dt) - i(t)f(t) - c\,dt] \qquad (5.2)$$

The optimal debt ratio is the *control variable* that maximizes the *criterion function*, the expected logarithm of net worth, subject to the *stochastic differential equation* (5.2). The main uncertainty is in the capital gain term dP(t)/P(t), which is unknown when the debt ratio is selected. I also allow uncertainty about the interest rate. The models discussed here differ in the assumptions concerning the stochastic process on the capital gains term. Hence there are differences in what is the optimal debt ratio. The optimal debt ratio was derived and explained in Chap. 4. The term (dP(t)/P(t) + β(t) dt) differs among models.

Models I and II are presented and discussed. I show in Chap. 6 that neither Model I nor Model II can be rejected by standard econometric tests. I then derive empirical measures of the optimal debt ratio appropriate for each model. The next step is to compare the actual f(t) to the derived optimal debt f*(t). In Sect. 5.5, I derive an upper bound f** > f* for both ratios to calculate the excess debt Ψ(t) = f(t) − f**(t), which is a warning signal of a 2007–2008 crisis. In Chap. 3 on the Quants, I explained why the actual debt ratio deviated from the optimal, and led to the crisis.

5.4.1 *Model I*

In Model I there are two sources of uncertainty: the price of the asset and the interest rate. Equations 5.3 and 5.4 concern the price P(t) of the asset. The productivity of capital $\beta(t) = \beta$ can be viewed as deterministic or constant and consumption ratio c(t) = c is constant. Model I contains two ideas, inspired by Bielecki and Pliska (1999) and Platen-Rebolledo (1995), and discussed in Fleming (1999). Equation 5.3 states that the price consists of two components, a trend and a deviation from it. The price trend is r. The initial value of the price is P = 1. The second component y(t) is a deviation from the trend. The second idea, expressed in (5.4) and (5.5), is that deviation y(t) is an ergodic mean reversion term whereby the price converges towards the trend. The speed of convergence of the deviation y(t) towards the trend is described by finite coefficient $\alpha > 0$. The stochastic term is $\sigma_p\, dw_p(t)$. The solution of stochastic differential equation (5.4) is (5.5). The deviation from trend converges to a distribution with a mean of zero and a variance of $\sigma^2/2\alpha$.

$$P(t) = P\,exp\,(rt + y(t)), \qquad P = 1, \tag{5.3}$$

$$y(t) = ln\,P(t) - ln\,P - rt. \tag{5.3a}$$

$$dy(t) = -\alpha y(t)dt + \sigma dw_p(t). \tag{5.4}$$

$$\begin{aligned} &\infty > \alpha > 0, \qquad \mathrm{E(dw)} = 0,\ \mathrm{E(dw)}^2 = \mathrm{dt}. \\ &lim\,y(t) \sim N(0, \sigma^2/\,2\alpha). \end{aligned} \tag{5.5}$$

The choice of price trend r is very important in determining the optimal leverage. I impose a constraint that the assumed price trend must not exceed the mean rate of interest. If this constraint is violated, as occurred during the housing price bubble, debtors were offered a "free lunch" as described in Chap. 3. Borrow/Refinance the house and incur a debt that grows at i, the mean rate of interest. Spend the money in any way that one chooses. Insofar as the house appreciates at a rate greater than the mean rate of interest, at the terminal date T the house is worth more than the value of the loan, P(T) > L(T). The debt L(T) is easily repaid by selling the house at P(T) or refinancing. One has had a free lunch. In the optimization, one must constrain the trend r not to exceed the *mean* rate of interest i(t). This constraint is (5.5).

$$r \leq i. \quad \text{No free lunch constraint} \tag{5.5}$$

An alternative justification for (5.5) is as follows. The present value PV of the asset

$$PV(T) = P(0)\,exp\,[(r - i)t], \tag{5.6}$$

where trend r is the rate of appreciation or capital gain and i is the mean interest rate. If (r − i) > 0, the present value diverges to plus infinity. An infinite present value is not sustainable. The real rate of interest is (i − r), which should be non-negative.

Equation 5.7 concerns the interest rate uncertainty. The interest rate i(t) is the sum of a constant i dt plus a Brownian Motion (BM) term $\sigma_i\ dw_i(t)$. The two BM terms, in (5.4) and (5.7) are correlated $0 > \rho \geq -1$, where the correlation ρ is presumed to be negative. When interest rates rise (decline) there are capital losses (gains). This negative correlation was an important factor in the house price bubble.

$$i(t) = i\,dt + \sigma_i\,dw_i(t). \tag{5.7}$$

The optimal debt ratio in Model I is (5.8), Box 5.1. The derivation is in Appendix Chap. 4. It states that the optimal debt ratio f*(t), that maximizes the expected growth of net worth, is positively related to the productivity of capital less the real rate of interest [β − (i − r)] = Net Return, and negatively related to the deviation of the price from trend αy(t) and negatively related to risk terms contained in σ^2 (5.8a). The negative correlation ρ < 0 between the capital gain and the interest rate increases the risk σ^2 and reduces the optimal ratio. Equation 5.8b expresses in words the content of (5.8) and (5.8a).

Box 5.1 Optimum Debt Ratio f*(t)

Model I

$$f^*(t) = \{[\beta - (i - r)] - (1/2)\,\sigma_p^2 - \alpha y(t) + \rho\sigma_i\sigma_p\}/\,\sigma^2 \quad (5.8)$$

$$\sigma^2 = \sigma_p{}^2 + \sigma_y{}^2 - 2\rho\sigma_i\sigma_p \qquad \rho<0 \quad (5.8a)$$

Net return N = $\{[\beta - (i - r)] - (1/2)\,\sigma_p{}^2 - \alpha y(t)\}$, RISK = σ^2

Optimal debt ratio = [Net Return − Risk elements]/Risk, (5.8b)

Model II

$$f^*(t) = f^* = [\beta + (\pi - i) - (\sigma_p{}^2 - \rho\sigma_i\sigma_p)]/\,\sigma^2 \quad (5.11)$$

$$\sigma^2 = \sigma_i{}^2 + \sigma_p{}^2 - 2\rho\sigma_i\sigma_p = \text{Risk} \quad (5.11b)$$

5.4.2 Model II

The stochastic variables in (5.2) and Model II are: (1) the capital gain dP(t)/P(t) and (2) the interest rate i(t). The capital gain has a drift of $\pi\ dt$ and a diffusion of $\sigma_p dw_p$ (5.9). The productivity of capital β(t) is deterministic or constant. The interest rate in (5.10) has a similar structure to (5.9). The mean is i dt and the BM term is $\sigma_i dw_i$ whose expectation is zero.

$$dP(t)\,/\,P(t) + \beta(t) = \pi\ dt + \beta(t)\ dt + \sigma_p dw_p \quad (5.9)$$

$$i(t) = i\,dt + \sigma_i dw_i \quad (5.10)$$

As in Model I, the two BM terms are expected to be negatively correlated, $\rho\ dt = E(w_b dw_i) < 0$. When the interest rate declines (rises), there are the capital gains (losses).

The optimal debt/net worth f*(t), derived in Chap. 4, is (5.11), where the optimal ratio is f*, a constant. The variance or risk is (5.11b) in Box 5.1. The NET RETURN = [π + (β − i)] is positively related to the productivity of capital plus *drift* of capital gain less the *mean* rate of interest. The debt ratio is negatively related to risk terms contained in σ^2 (5.11b). The negative correlation $\rho < 0$ between the capital gain and the interest rate increases the risk σ^2 (5.11b) and reduces the optimal ratio.

5.5 Interpretation of Optimal Debt Ratio

Figure 4.3 describes the optimal debt ratio f*(t) in a manner that is valid for either model. The optimal debt ratio is on the vertical axis. It is (5.8) for Model I and (5.11) for Model II. The horizontal axis measures the Net Return. The optimal debt ratio is a linear function of the Net Return. The risk premium is R in Fig. 4.3. The optimal debt ratio is positive only if the Net Return exceeds the risk premium.

The SOC analysis in Chap. 4 proved that the expected growth of net worth is maximal along the "optimal debt ratio line". This line relates the optimal debt ratio to the net return. The optimal debt ratio is not a constant, but varies directly with the net return and risk. If the Net Return is N(1), the optimal debt ratio is f*(1). As the actual debt ratio f(1) rises above the optimal debt ratio the line, the expected growth declines and the risk rises. If the actual debt ratio is f(1) and the net return is N(1), there is an excess debt $\Psi(t) = f(1) - f^*(1) > 0$ and the expected growth is less than maximal. The larger is the excess debt, the lower the expected growth, the greater is the variance and hence the greater the probability that net worth will be driven towards zero. An early warning signal of a crisis is an excess debt. *The main conclusion is that it is not the debt per se that leads to a crisis, but it is the excess debt.*

5.6 Empirical Measures of an Upper Bound of the Optimal and Actual Debt Ratio

The *optimum debt ratio* f* is based upon the equations in Box 5.1, with the constraint that the *mean* real interest rate $(i - r)$ is non negative – to avoid the "free lunch" difficulty. There are other sensible stochastic processes, discussed in Chap. 4. No one knows what is the correct way to model the stochastic processes. My strategy is to derive an upper bound for the optimal debt ratio f*(t) in Box 5.1 which will also be compatible with alternative sensible stochastic processes. Call the upper bound ratio f**. Then the excess debt $\Psi(t) = f(t) - f^{**}(t)$ is measured as the actual debt ratio less an upper bound f** ratio. The most problematic issue is how to measure the stochastic price, the capital gain dP(t)/P(t).

From the histogram of the capital gains in Fig. 2.1, the mean capital gain over the period 1980q1–2007q4 was 5.4% per annum with a standard deviation of 2.9%. The 30-year conventional mortgage rate in Fig. 3.3 ranged from 7.4% to 5% pa from 2002 to 2007. It is reasonable to argue that, over a long period, the appreciation of housing prices was not significantly different from "the mean mortgage rate of interest". Define $(i - r)$ as the mean interest rate less the trend growth of price, called the mean real rate of interest. A value of $(i - r) \geq 0$, a non-negative mean real rate of interest precludes the "free lunch" that characterized the bubble. Then one component of an upper bound f**(t) in Model I equation (5.8) for the optimal debt ratio is (i – r) = 0. The other variable is y(t) the deviation of price from trend. Since it is difficult to estimate the appropriate trend or price drift, for reasons discussed in Chap. 3, I assume that deviation from trend y(t) = 0. Using these two assumptions in (5.8) an upper bound of the optimal debt ratio f**(t) is (5.12). We take *β(t)* to be deterministic or a constant.

$$f^{**}(t) = \{[\beta(t) - (1/2)\,\sigma_p{}^2 + \rho\sigma_i\sigma_p\}/\sigma^2 \qquad y(t) = 0,\ (i - r) = 0 \tag{5.12}$$

In Model II, the corresponding assumption is that the mean drift of the price is equal to the mean rate of interest, π = i. Then f**(t) for Model III is (5.13).

$$\mathrm{f^{**}(t)} = [\beta(\mathrm{t}) - (\ \sigma_\mathrm{p}{}^2 -\ \rho\sigma_\mathrm{i}\sigma_\mathrm{p})]/\sigma^2 \tag{5.13}$$

$$f^{**}(t) = [(R(t)\,/\,P(t)) - (risk\ elements)]\,/\,\sigma^2 {>} f^*(t)$$
$$Mean f^{**}(t) = [mean\,R(t)/P(t) - (risk\ elements)]/\sigma^2 \tag{5.13a}$$

We must estimate β(t), the productivity of capital. The productivity of housing capital is the implicit net rental income/value of the home plus a convenience yield in owning one's home. Assume that the convenience yield in owning a home has been relatively constant. The productivity of capital is rental income R(t) divided by the value of housing P(t)Q(t), where P(t) is an index of house prices and Q(t) is an index of the "quantity" of housing. Thus β (t) = R(t)/P(t)Q(t). I approximate the return β(t) by using the ratio R(t)/P(t) of rental income/an index of house prices. This is (5.13), whose mean is (5.13a).

The units of R(t) and P(t) are different. The numerator is in dollars and the denominator is a price index. In order to make alterative measures of the debt ratio and key economic variables comparable, I use normalized variables where the normalization (N) of a variable Z(t) called N(Z) = [Z(t) – mean Z]/standard deviation. The normalized f**(t) is basically (5.13) minus (5.13a). The risk elements in the numerator cancel.

The mean of N(Z) is zero and its standard deviation is unity. The normalized upper bound f**(t) of optimal debt ratio is (5.14) or (5.14a). Call (5.14b) the RENTPRICE. It is the (rental income/index of house prices – mean)/standard deviation. The risk elements in the numerator of (5.13) cancel, in deriving the normalized value. The RENTPRICE graphed in Fig. 5.3 corresponds to an upper bound f**(t) of the optimal debt ratio.

$$[f^{**}(t) - mean f^{**}(t)] = [R(t)/P(t) - mean.]/\sigma^2 \tag{5.14}$$

$$N(f^{**}(t)) = [(\beta(t) - \beta)]/\sigma(\beta) \tag{5.14a}$$

$$N(f^{**}(t)) = \mathrm{RENTPRICE} \tag{5.14b}$$

Fig. 5.3 N[f(t)] = DEBTSERVICE = (household debt service payments as percent of disposable income i(t)L(t)/Y(t) − mean)/standard deviation. N[f**(t)] = RENTPRICE = (rental income/index house price R(t)/P(t) − mean)/standard deviation (Sources: FRED, OFHEO)

There are a couple of measures of the debt ratio. One measure of the actual debt ratio is L(t)/Y(t) household debt as percent of disposable income. This is series FODSP in FRED. A more useful measure is the debt service ratio i(t)L(t)/Y(t) which measures the *debt burden* graphed in Fig. 5.1. This is series TDSP in FRED. The debt L(t) includes all household debt, not just the mortgage debt, because the capital gains led to a general rise in consumption and debt. The *normalized value* of the debt ratio N(f) is (5.15), which is graphed in Fig. 5.3 as DEBTSERVICE. This is measured in units of standard deviations from the mean of zero. There is a dramatic deviation above the mean from 1998 to 2005. This sharp rise coincides with the ratio P/Y of housing price index/disposable income, P/Y = PRICEINC in Fig. 2.2. During this period, there was more than a two standard deviation rise in P/Y and a two standard deviation rise in household debt/disposable income.

$$N(f) = DEBTSERVICE = [i(t)L(t) / Y(t) - mean] / standard\ deviation. \quad (5.15)$$

Compare the DEBT SERVICE with the RENT PRICE in Fig. 5.3. This deviation will be used as an upper bound on excess debt.

5.7 Early Warning Signals of the Crisis

My argument in connection with Figs. 4.2 and 4.3 is that the excess debt Ψ(t) = f(t) − f**(t) is an early warning signal of a crisis. The next question is how to estimate the excess debt Ψ(t) that corresponds to the models in Box 5.1 and is consistent with alternative estimates of the optimal debt.

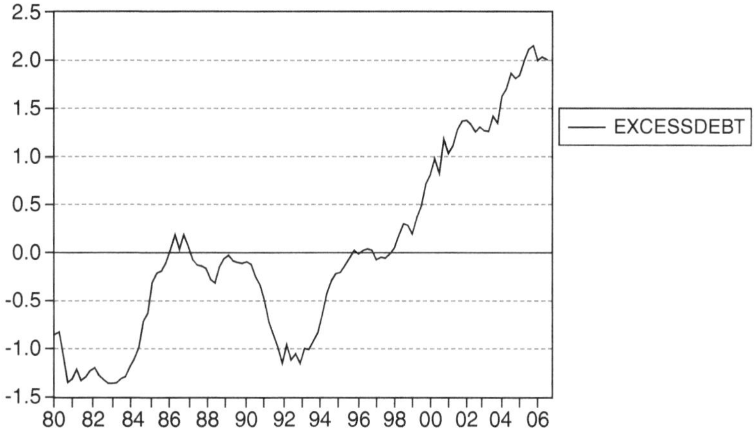

Fig. 5.4 Early Warning Signal since 2004. Ψ(t) = EXCESS DEBT = N[f(t)] − N[f**(t)] = DEBTSERVICE i(t)L(t)/Y(t) − RENTRATIO R(t)/Y(t)

I estimate excess debt Ψ(t) = (f(t) − f**(t)) by using the difference between two normalized variables N(f) − N(f**), in Fig. 5.3. This difference is *measured in standard deviations*. Debt service is i(t)L(t)/Y(t) household debt service payments as percent of disposable income, Fig. 5.1 above, (TDSP in Federal Reserve St. Louis FRED).

$$\begin{aligned}\Psi(t) &= EXCESS\,DEBT = N[f(t)] - N[f^{**}(t)] \\ &= Normalized\,\{DEBTSERVICE\; i(t)L(t)/Y(t) - RENTPRICE\; R(t)/P(t)\}.\end{aligned} \tag{5.15a}$$

Excess Debt graphed in Fig. 5.4 corresponds to the difference Ψ(t) = f (t) − f**(t) on the vertical axis in Fig. 4.3, measured in standard deviations. The SOC analysis in Chap. 4 proved that the probability of a decline in net worth is positively related to Ψ(t) the excess debt. As the excess debt rises, the expected growth declines and the risk increases.

Assume that over the entire period 1980–2007 the debt ratio was not excessive. There was no excess debt before 1999. During the period 2000–2004, the high capital gains and low interest rates induced rises in housing prices relative to disposable income and led to rises in the debt service ratio. By 2004–2005 the ratio of housing price/disposable income was about three standard deviations above the long-term mean. See PRICEINC in Fig. 2.2, Chap. 2. This drastic rise alarmed several economists who believed that the housing market was drastically overvalued. As indicated in Chap. 2, they were in a minority. It certainly had a negligible effect upon the market for derivatives and the optimism of the “Quants”.

The advantages of using excess debt Ψ(t) in Fig. 5.4 as an Early Warning Signal compared to just the ratio of housing price/disposable income are that Ψ(t) focuses upon the fundamental determinants of the optimal debt ratio as well as upon the actual ratio. The excess debt reflects the difference between the debt obligations and the ability to service them. The probability of declines in net worth, the inability of the mortgagors to service their debts and the financial collapse due to leverage – a crisis – are directly related to the excess debt.

Based upon Fig. 5.4, early warning signals were given as early as 2003. By 2005, the excess debt was about two standard deviations above the mean. Excess debt is not normally distributed, as is obvious from Fig. 5.4. A probability cannot be assigned in this case. But whatever the distribution, a 2-3 standard deviation from the mean must be viewed as an Early Warning Signal (EWS). The actual debt was induced by capital gains in excess of the mean interest rate. The debt could only be serviced from capital gains. This situation is unsustainable. When the capital gains fell below the interest rate, the debts could not be serviced. A crisis was inevitable.

The magnitude of the excess debt in Fig. 5.4 is a benchmark to evaluate that the economy is being carried away in a bubble. Greenspan, the Fed and IMF had no concept of this and were oblivious to the EWS since 2004.

5.8 The Market Delusion

A crucial variable in determining the "optimal" debt/net worth in Model I for example is the trend of the housing price. Figure 5.5 describes the house price appreciation HPI or capital gains variable CAPGAIN = [P(t) − P(t − 1)]/P(t) denoted by $\pi(t)$, which is the four-quarter appreciation of US housing prices, sample 1991q1–2011q1.

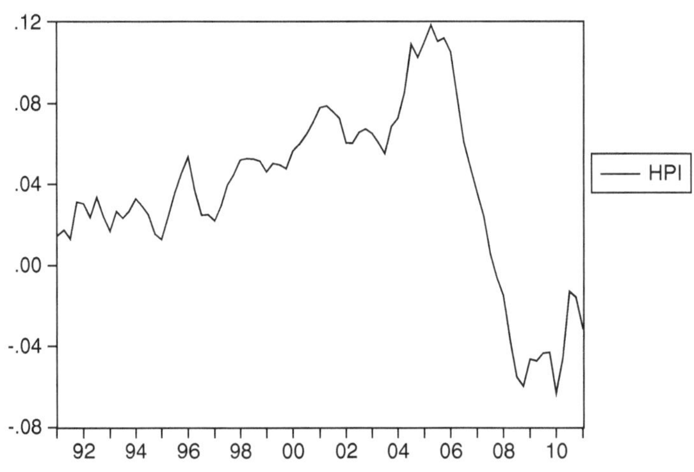

Fig. 5.5 Capital gain CAPGAIN = [P(t) − P(t − 1)]/P(t). House Price Index, change over previous four quarters (e.g. 0.08 = 8% pa) (Federal Housing Finance Industry FHFA USA Indexes. Sample: 1991q1–2011q1)

In Chap. 3 I explained how the Quants/Market estimated the trend of the capital gain, with disastrous results. They assumed that the recent capital gains 1994–2004 could continue. These values exceeded the mean rate of interest and seemed to offer a "free lunch". My analysis above stresses that optimization should be based upon the assumption that the mean interest rate must exceed the longer run capital gain. This was violated from 1998 to 2004. Many were aware of that but believed that one could get out "as soon as the music stopped". This was not possible since they all used the same models and had the same data. All could not get out at the same time.

5.9 The Shadow Banking System: Leverage and Financial Linkages

The financial crisis was precipitated by the mortgage crisis for several reasons. First, a whole structure of financial derivatives was based upon the ultimate debtors – the mortgagors. Insofar as the mortgagors were unable to service their debts, the values of the derivatives fell. Second, the financial intermediaries whose assets and liabilities were based upon the value of derivatives were very highly leveraged. Changes in the values of their net worth were large multiples of changes in asset values. Third, the financial intermediaries were closely linked – the assets of one group were liabilities of another. A cascade was precipitated by the mortgage defaults. Since the "Quants" were following the same rules, the markets would not be liquid.

Charles Prince (former Citigroup CEO) told the FCIC that: "Securitization could be seen as a factory line. As more and more of these subprime mortgages were created as raw material for the securitization process, not surprisingly in hindsight, more and more of it was of lower and lower quality. At the end of that process, the raw material going into it was actually of bad quality, it was toxic quality, and that is what ended up coming out the other end of the pipeline. Wall Street obviously participated in that flow of activity".

A summary of the linkages in the financial system, discussed below, is sketched in Box 5.2. My discussion of the role of the shadow banking system is directly based upon the FCIC Report. See also the discussion in Chap. 3 on the "Quants". The next chapter is devoted to the AIG case. The origination and securitization of the mortgages relied upon short-term financing from the shadow banking system. Unlike banks and thrifts with access to deposits, investment banks relied more on money market funds and other investors for cash. Commercial paper and repo loans were the main sources. This flood of money and securitization apparently helped boost home prices from the beginning of 2004 until the peak in April 2005, even though homeownership was falling.

Box 5.2 Leverages and Linkages in the Shadow Banking System

Households/Mortgagors = > Originator → *Shadow Banking System*

(a) Major banks, securitization firms, investment banks: Bear Stearns – vertical integration model, Merrill Lynch, Lehman, Morgan Stanley, Goldman-Sachs, Citigroup, hedge funds. Repackaging of mortgages for sale to investors.
(b) Short-term financing from banks and Money Market Funds: secured by mortgages. Repo, commercial paper

Vulnerabilty of shadow banking system: high leverage, short-term financing, collateral calls, inadequate liquidity. Inadequate information about risks.

i. Mark-to-market reflects House price index (HPI). Decline in HPI reduces value of collateral and of net worth.
ii. Contagion. Lenders question value of assets of financial firms and of the collateral posted. Require more/safer collateral. Counterparty risks.
iii. Risk concentrated. High leverage ratios of investment banks up to 40:1, inadequate capital, short term funding, leveraged with commercial paper, repos – using CDOs, Mortgage Backed Assets as collateral.

Leverage (assets/net worth): Citigroup 18:1 (2000) to 32:1 (2007); G-S 17:1 (2000) to 32:1 (2007); Morgan Stanley, Lehman 40:1 (2007).

Revenues, earnings from trading, investing securitization, derivatives:

(a) Revenues: G-S 39% (1997) to 58% (2007); M-L; 42% (1997) to 55% (2005);
(b) Pretax earnings: Lehman 32% (1997) to 80% (2005); Bear Stearns 100% in some years after 2002.

The shadow banking system consisted primarily of investment banks, but also other financial institutions – that operated freely in capital markets beyond the reach of regulatory agencies. Among the largest buyers of commercial paper and repos were the money market mutual funds. Commercial paper was unsecured corporate debt, short-term loans that were rolled over. The market for repos is a market for loans that are collateralized. It seemed like a win-win situation. Mutual funds could earn solid returns and stable companies could borrow more cheaply, Wall St. firms could earn fees putting the deal together. The new parallel banking system, with repos and commercial paper providing better returns for consumers and institutional investors, came at the expense of banks and thrifts.

Over time, investment banks and securities firms used securitization to mimic banking activities outside the regulatory framework. For example, whereas banks traditionally took money from deposits to make loans and held them to maturity, the investment banks used money from the capital market, often from Money Market Mutual Funds, to make loans, package them into securities to sell to investors.

From 2000 to 2007, large banks generally had leverage (assets/net worth) ratios from 16:1 and then up to 32:1 by the end of 2007. Because investment banks were not subject to the same capital requirements as commercial and retail banks, they were given greater latitude to rely on internal risk models in determining capital requirements, and reported higher leverage. At Goldman-Sachs, leverage increased from 17:1 (2001) to 32:1 (2007). Morgan Stanley and Lehman reached 40:1 at the end of 2007. Trading and investments, including securitization and derivative activities, generated an increasing amount of investment banks' revenues and earnings. At G-S, revenues from trading and principal investments increased from 39% (1997) to 58% (2007). At Merrill Lynch revenue from those activities rose from 42% (1997) to 55% (2005). At Lehman, similar activities generated 80% of pre-tax earnings in 2005 up from 32% in 1997. At Bear Stearns, they accounted for 100% of pre-tax earnings in some years after 2002.

Foreign investors sought the high-grade securities but at a higher return than obtained on US Treasuries. They found triple-A assets flowing from the Wall St. securitization machine. As overseas demand drove up prices of securitized debt, it created an irresistible profit opportunity for the US financial system to engineer "quasi-safe" debt instruments by bundling riskier assets and selling senior tranches. See the discussion in Chap. 3. The US housing bubble was financed by large capital inflows, as occurred in Ireland and Spain discussed in a later chapter.

Some of the key players in the financial network were Lehman, Citigroup, Bear Stearns and Merrill Lynch. Since the late 1990s *Lehman* had built a large mortgage arm, a formidable securities issuance business and a powerful underwriting as well. Lehman moved to a more aggressive growth including greater risk and more leverage. It moved to a "storage" business where Lehman would make and hold longer-term investments. Across both the commercial and residential sectors, the mortgage related assets on Lehman's books increased from $52 billion (2005) to $111 billion (2007). This would lead to Lehman's undoing a year later. Lehman's regulators were aware of this but did not restrain its activities.

As late as July 2007, *Citigroup* was still increasing its leveraged loan business. Citigroup broke new ground in the CDO market. It retained significant exposure to potential losses on its CDO business, particularly with Citibank – the $1 trillion commercial bank whose deposits were insured by the FDIC. In 2005 Citigroup retained the senior and triple-A tranche of most of the CDOs it created. In many cases, Citigroup obtained protection from a monoline insurer. Because these hedges were in place, Citigroup presumed that the risk associated with the retained tranches had been neutralized.

Citigroup reported these tranches at values for which they could not be sold, raising questions about the accuracy of reported earnings. Part of the reason for retaining exposures to super-senior positions in CDOs was their favorable capital treatment. Under the 2001 Recourse Rule, banks were allowed to use their own models to determine how much capital to hold, an amount that varied according to how much the market prices moved. Citigroup judged that the capital requirements for the super-senior tranches of synthetic CDOs it held for trading purposes was effectively zero, because the prices did not move much. As a result, Citigroup held little regulatory capital against the super-senior tranche.

Merrill Lynch focused on the CDO business to boost revenue. To keep its CDO business going, M-L pursued several strategies, all of which involved repackaging riskier mortgages more attractively or buying its own products, when no one else would like to do it. Like Citigroup, M-L increasingly retained for its own portfolio substantial portions of the CDOs it was creating, mainly the super-senior tranches and it increasingly repackaged the hard to sell BBB rated and other low rated tranches into other CDOs. Merrill-Lynch continued to push its CDO business despite signals that the market was weakening.

The bust showed the weaknesses of the system based upon the homeowners' mortgage debt. Early 2007, it became obvious that home prices were falling in regions that had once boomed, and that mortgage originators were floundering and that more and more would be unable to make mortgage payments. *The question was how the housing price collapse would affect the financial system that helped inflate the bubble.* In theory – held by Greenspan, Bernanke and the market – securitization, over the counter derivatives and the shadow banking system was supposed to distribute risk efficiently among investors. In fact, much of the risk of mortgage backed securities had been taken by a small group of systemically important companies, with huge exposures to senior triple-A tranches of CDOs that supposedly were "super-safe".

As 2007 went on, increasing mortgage delinquencies and defaults compelled the rating agencies to downgrade the mortgage backed securities and then the CDOs. Mortgage delinquencies had hovered around 1% during the early part of the decade jumped in 2005 and kept climbing to 9.7% at the end of 2009. Alarmed investors sent CDO prices plummeting. Hedge funds faced with margin calls from their repo lenders were then forced to sell at distressed prices. Many would shut down. Banks wrote down the value of their holdings by billions of dollars.

By the end of 2008, more than 90% of all tranches of the CDOs had been downgraded. Moody's downgraded nearly all of the Baa CDO tranches. The downgrades were large. More than 80% of Aaa and all of the Baa CDO bondswere eventually downgraded to junk. As market prices of CDOs dropped, *"Mark-to-Market" (M-t-M)* accounting rules require firms to write down their holdings to reflect the lower market prices. The M-t-M write downs were required on many securities even if there were no actual realized losses, and in some case, even if the firms did not intend to sell the securities.

Determining the market value of securities that did not trade was difficult and subjective and became a contentious issue during the crisis, because the write-downs reduced earning and triggered collateral calls. Large, systemically important firms with significant exposure to these securities would be found to be holding very little capital to protect against potential losses. See Chap. 6 on AIG. Most of these companies would turn out to be considered by the authorities as "too big to fail" in the midst of the financial crisis.

When the crises began, *uncertainty and leverage would promote contagion.* Investors realized that they had limited information about the mortgage assets that banks and investment banks held and to which they were exposed. Financial institutions had leveraged themselves with commercial paper, with derivatives and in the short term repo market, in part using CDOs as collateral. Lenders questioned the value of assets that these companies had posted as collateral at the same time that they were questioning the value of those companies' balance sheets.

Bear Stearns (B-S) hedge funds had significant subprime exposure and were affected by the collapse of the housing bubble. The collapse produced pressure on the parent company. The commercial paper and repo markets – the two key components of the shadow banking lending market – quickly reflected the impact of the housing bubble collapse, because of the decline of the value of collateral and concern about the firms' subprime exposure.

In July 2007, B-S hedge funds failed. Its repo lenders – mostly money market mutual funds – required that B-S post more collateral and pay higher interest rates. Mortgage securitization was the biggest piece of B-S most profitable portfolio division. In mortgage securitization, B-S followed a vertically integrated model that made money at every step, from loan origination through securitization and sale. It both acquired and created its own captive organization from originating and securitization/bundling of mortgages into securities that were then sold to investors.

Nearly all hedge funds provide their investors with market value reports, based upon computed M-t-M prices for the fund's various investments. For mortgage backed investments, M-t-M assets was an extremely important exercise, because market values (1) were used to inform investors, (2) to calculate the hedge fund's total value for internal risk management purposes and (3) because these assets were held as collateral for repo and other lenders. The crux was that if the value of a hedge fund's portfolio declined, repo and other lenders might require more collateral. Many repo lenders sharpened their focus on the valuation of any collateral with potential subprime exposure. Lenders required increased margins on loans to institutions that appeared to be exposed to the mortgage market. They often required Treasury securities as collateral. Triple-A rated mortgage-backed securities were no longer considered super-safe investments.

Bankers and regulators knew that other investment banks shared B-S weaknesses: leverage, reliance on overnight funding, concentration in illiquid mortgages. As Bear Stearns hedge funds were collapsing, the market viewed Bear Stearns as "the canary in the mine shaft". During all of this Bernanke was not unduly unperturbed. In his testimony to the Joint Economic Committee March 28, 2007, he said: "The impact on the broader economy and financial markets of the problems in the subprime market seems likely to be contained".

Lehman Brothers was another key player and here is where solvency and liquidity interacted. If assets are greater than liabilities, net worth (capital) is positive and the firm is solvent. If not, the firm is in danger of bankruptcy. In August 2008, shareholder equity was $28 billion. However, balance sheet net worth (capital) is not relevant if the firm is suffering a massive run. If there is a run, and a firm can only get fire sale prices for assets, even large amounts of capital can disappear almost overnight. The chief concerns were Lehman's real estate related investments and its reliance on short term funding, including $7.8 billion of commercial paper and repos as of the end of 2008q1. People were demanding liquidity from Lehman, which it did not have.

The most pressing danger was the potential failure of the repo market. The little regulated repo market grew from an average daily volume $800 billion (2002) to $2.8 trillion (2008). It had become a very deep and liquid market. Even though most borrowers rolled over repos overnight, it was considered a very safe market because transactions were over collateralized, the ratio of loans/collateral was less than unity. Even long term repo loans had to be unwound every day. To the surprise of both lenders and regulators, as a result to the failure of B-S hedge funds, high quality collateral was not enough to ensure access to the repo market.

If lenders had to sell off large amounts of collateral in order to meet their own cash needs, that action would lead to widespread fire sales of repo collateral and runs by lenders. Most of the Over the Counter derivatives dealers hedge their contracts with offsetting contracts, creating the potential for a series of losses and defaults. This was a *systemic risk* to the financial system as a whole.

In March 2008 the Fed provided a loan to facilitate JP Morgan's purchase of B-S. Bernanke refused to lend to Lehman, because Lehman did not have sufficient collateral. This inconsistency has been the subject of intense debate. See Chap. 6 for the AIG case.

5.9.1 Summary

One may summarize the crisis in the following way. Chairman Greenspan injects reserves/lowers short-term interest rates in response to the bursting dot-com bubble. Mortgage rates fall spurring refinances and speculative interest in the housing market. House prices start rising, reducing current Loan/Values and creating perception of reduced credit risk in mortgage lending. Banks securitize mortgage loans into mortgage backed securities MBS which in turn are resecuritized into CDOs, and are met with astounding investor demand from yield-hungry institutions. As investor demand drives down securitization costs, banks reduce mortgage yields and loosen credit standards to entice more borrowers into the market. These borrowers in turn drive housing demand upward leading to high rates of House Price Appreciation which in turn lower perceived probabilites of losses and therefore reduce perceived credit risk and risk premia required by MBS investors. Banks continue to earn high fees and spreads by originating and then selling mortgage risk off their balance sheets, rating agencies continue to earn high fees from rating MBS and CDOs, securities firms earn their bid/offer spreads, and real estate speculators continue to flip houses and mortgagors live well above their means enjoying their free lunches. Then, enough people start realizing that the house price to rental ratio is "too high", that the house price to median income ratio is "too high" and as soon as house prices level off (not even start to fall) borrowers who never could afford the debt service costs are caught in a situation where they can no longer sell and pay off their loans at a profit or refinance based on the capital appreciation of the property. Then the cascade downward begins.

My conclusions are consistent with those of Juselius–Kim. "When is financial stability jeopardized by systemic events? Is there a maximum sustainable debt burden. . .? If there is, can it be empirically estimated and will exceeding it necessarily result in a crisis? These are timely and important questions in the wake of the recent financial crisis and its evolution, yet empirical evidence on these issues is rather scarce. Using data on aggregate US credit losses, we show that debt sustainability depends on financial obligations ratios (interest payments and amortizations relative to income) . . .. In particular, we are able to estimate threshold values, interpreted as maximum sustainable debt burdens, for the financial stability is jeopardized by systemic events." I differ from them insofar as it was the excess debt that led to the crisis, whose severity was aggravated by the leverages in the financial sector.

References

Bielecki, T.R., and S. Pliska. 1999. Risk sensitive dynamic risk management. *Applied Mathematical Optimization* 39: 337–350.

Congressional Oversight Panel (COP). 2011. Report on troubled assets relief program, March. Washington, DC.

Congressional Oversight Panel (COP). 2009. Special report on regulatory reform, Washington, DC.

Demyank, Y., and O. Van Hemert. 2008. Understanding the subprime mortgage crisis, SSRN working paper.

Federal Housing Finance Agency. House Price Index USA OFHEO House Price Index http://www.fhfa.gov.

Federal Reserve Bank, St. Louis. FRED data bank.

Financial Crisis Inquiry Report, FCIC. 2011. *Public affairs*. New York.

Fleming, W.H. 1999. Controlled Markov processes and mathematical finance. In *Nonlinear analysis, differential equations and control*, ed. F.H. Clarke and R.J. Stern. Dordrecht: Kluwer.

Juselius, M., and M. Kim 2011. Sustainable financial obligations and crisis cycles, HECER discussion paper 313.

Office of Federal Housing Price Oversight (OFHEO).

Platen, E., and R. Rebolledo. 1995. Principles for modeling financial markets. *Journal of Applied Probability* 33: 501–530.

Chapter 6
AIG in the Crisis

Abstract The financial structure of the derivatives insured by AIG with credit default swaps (CDS) was ultimately related to the systemic risk from the inability of the mortgagors to service their debts. AIG made several serious mistakes. First: The estimate of the drift of the capital gain, which drove the bubble, was based upon the unsustainable growth of the housing price index 2004–2006. A collapse would occur when the unsustainable capital gain declined below the interest rate. Second, risk was underestimated because AIG ignored the negative correlation between the capital gains and the liabilities/claims. The CDS claims grew when the value of the insured obligations declined. This set off collateral requirements, and the stability of AIG was undermined. The solution for the optimal insurance liabilities on the basis of SOC is derived. The SOC approach is a generalization of the contributions of the economics and actuarial literature. The chapter concludes with an evaluation of the government bailout.

6.1 Introduction

At its peak, American International Group (AIG) was one of the largest and most successful companies in the world boasting a Triple-A credit rating, over $1 trillion in assets, and 76 million customers in more than 130 countries. Yet the sophistication of AIG's operations was not matched by an equally sophisticated risk-management structure. This poor management structure, combined with a lack of regulatory oversight, led AIG to accumulate staggering amounts of risk, especially in its subsidiary, AIG Financial Products (AIGFP). I draw upon the Congressional Oversight Panel Report (COP 2010) that describes AIG's role in the financial market crisis.

Section 6.1 describes what happened to AIG in the 2007–2008 crisis. Section 6.2 sketches the economics literature and the actuarial literature. The conclusions from the economics literature are that "Despite extensive researches about the determinants of corporate capital structure, the theories of capital structure remain

J.L. Stein, *Stochastic Optimal Control and the U.S. Financial Debt Crisis*,
DOI 10.1007/978-1-4614-3079-7_6, © Springer Science+Business Media New York 2012

one of the most controversial issues in modern corporate finance" (Shim 2010). The actuarial literature on capital requirements and risk was discussed in Chap. 4. The stochastic optimal (SOC) approach in the following sections is a generalization of the contributions of the economics and actuarial literature. Its components are: the criterion function, the stochastic differential equations, and the stochastic processes. The solution for the optimal capital requirement on the basis of SOC is in Sect. 6.5. Section 6.6 is an evaluation of the government bailout.

6.2 AIG

AIG occupied an important role in the financial system. Of all the possible losers in the financial crisis discussed in Chap. 5, AIG should have been the most concerned with risk management. The AIG Financial Products AIGFP subsidiary sold credit default swaps (CDSs) involving three parties. Party 1 the obligor sells the "reference securities" to party 2. The latter are security houses, hedge funds that purchase protection from party 3, who are banks and insurance companies, which issue the protection. In many cases municipalities are required to carry insurance in order to market their bonds, or to obtain a high credit rating. The CDS are privately negotiated contracts that function in a similar manner to insurance contacts, but their payoff structure is closer to a put option.

AIGFP's obligations were guaranteed by AIG's triple-A rating, which facilitated borrowing at low interest rates. Ultimately, however, it made it difficult to isolate AIGFP from its parent, with disastrous consequences. Bernanke characterized AIGFP as a hedge fund attached to a large and stable insurance company. AIGFP entered the credit derivatives market in 1998 when it underwrote its first CDS.

Over time AIGFP became a central player in the fast-growing CDS market.

AIGFP's corporate arbitrage CDS portfolio was comprised of CDS contracts written on corporate debt and collateralized loan obligations (CLOs). The collateral pools backing the corporate debt portfolio included baskets of investment-grade corporate bonds and loans of commercial and industrial loans of large banks. The collateral pools backing the multi-sector CDOs included prime, Alt-A, and subprime residential mortgage-backed securities (RMBS), commercial mortgage-backed securities (CMBS), other asset-backed securities (ABS). CDS written on corporate debt, CLOs, and multi-sector CDOs serve as protection against "credit events" of the issuer of the reference obligation, including bankruptcy, failure to pay, acceleration of payments on the issuer's obligations, default on the issuer's obligations, restructuring of the issuer's debt, and similar events.

AIGFP's operating income grew from \$131 million in 1994 to \$949 million in 2006, 18% pa, paralleling the boom in the overall derivatives market, as well as the CDS market. While the credit markets provided a source of steady profits for AIGFP, the division's operating income represented a relatively small percentage of AIG's total operating income, contributing just 7% to firm-wide net income in 2006. More importantly, as events made clear, the risk involved in this business was dramatically disproportionate to the revenue produced. For example, *losses in 2007 totaled $11.5 billion, twice the aggregate net income produced by this division from 1994 to 2006.*

AIG models showed that there would be no defaults on any of the bond payments that AIG swaps had insured. AIGFP did not have its own model or otherwise try to value the CDO portfolio that it guaranteed through CDS nor did it hedge its exposure. AIG mistakenly believed that it would only have to pay counterparties if holders of the super-senior tranches insured incurred actual losses. AIGFP relied on Gorton's actuarial model that did not provide a tool for monitoring the CDO's market value. Gorton's model had determined with 99.85% confidence that the owners of super-senior tranches of the CDO's insured by AIGFP would never suffer real economic losses (FCIC, pp. 266–267). The company's auditor PWC apparently was also not aware of the collateral requirements. PWC concluded that "..the risk of default on the [AIG] portfolio has been effectively removed as a result of a risk management perspective...".

The bankruptcy of Lehman Brothers was a cataclysmic event. Investors began to compare the types of securities held by AIG and by Lehman. Goldman-Sachs was an AIG counterparty, which was very much concerned with the value of AIG's collateral. The G-S executives concluded that the Gorton/AIG models were irrelevant because the contracts required collateral to be posted if market values of the insured securities declined, irrespective of only long term cash losses. G-S estimated that the average decline in the market value of the bonds was 15%. With leverage this would translate into very large declines in equity.

Until the dispute with G-S, AIG relied on Gorton's model that did not estimate the market value of the underlying securities. So G-S mark to market caught AIG by surprise. By August 2007 AIG publicly disclosed, for the first time, the $79 billion in CDS's on super-senior CDO tranches. Sixty four billion were backed by subprime and nineteen billion were written on risky BBB collateral. By January 2008, AIG still did not have a reliable way to determine the market price of the securities on which it had written credit protection.

AIG valued its Alt-A and subprime mortgage backed securities at 1.7–2 times the values used by Lehman. During the summer of 2008, AIG faced increasing demands from their CDS customers for cash security – collateral calls – totaling tens of billions of dollars. AIG was not able to provide the required additional collateral. This put AIG's credit rating under pressure, which in turn led to even greater collateral calls, creating even greater pressure on AIG's credit.

In September 2008, AIG suffered a liquidity crisis, because the firm lacked the liquidity to meet collateral demands from its customers. In only a matter of months AIG's worldwide empire had collapsed, brought down by the company's underestimation of risk. The AIG liquidity crisis led to the largest government bailout of a private company in US history, totaling $182 billion. Maurice Greenberg, the former CEO, characterized the bailout as a nationalization of AIG, was bewildered by the situation and was at a loss over how the entire situation got out of control as it did.

Table 6.1 AIG and the financial crisis, fall 2008

Date 2008	(1) TED spread (bps)	(2) 3-month LIBOR – OIS spread (bps)	(3) AIG CDS spread (bps)	(4) AIG stock price ($)	(5) 3-month treasury bond yield (%)	(6) D-J industrial average
Aug. 15	96	77	300.7	459.8	1.85	11,659
Sept. 15	180	105	1,527.6	95.2	1.02	10,917.5
Oct. 15	433	345	1,816.9	48.6	0.22	8,577.9
Nov. 7	198	176	2,923.9	42.2	0.31	8,943.8

Source: COP (2010, Fig. 14). (1) Interest rate on interbank loans – US Treasury bill rate; (2) 3-month LIBOR – OIS overnight index spread rate; (3) A CDS rate of y basis points (bps) means that it costs $1000y to insure $10 million of debt for 5 years

The resulting downgrades of AIG securities were, in the opinion of Charles Prince, the precipitating event in the financial crisis. The Office of Thrift Supervision (OTS) was the only regulator that had explicit authority to look at the entire company, and the only regulator with any authority over AIGFP. Acting Director Polakoff acknowledged that AIG did not foresee the extent of risk concentration and profound systemic impact CDS caused within AIG. He stated that OTS should have directed AIG to stop originating CDSs and begin reducing its CDS portfolio before December 2005.

Table 6.1 shows that the crisis of both AIG and of the financial sector was *unexpected by the market and pervasive*. Column 1 is the spread TED between the interest rate on interbank loans and the US Treasury bill rate. Column 2 is 3-month LIBOR – OIS overnight index spread rate. It is a measure of how likely borrowing banks will default. Column 3 is a measure of the cost of insurance, based upon CDS.

In August 2008 markets were relatively tranquil. A crisis was not anticipated. One month later the crisis appeared. The cost of CDS insurance (column 3) was five times higher and the stock price of AIG fell by 80%. The financial crisis reached panic proportions in October. Banks were reluctant to lend to each other, as seen in the drastic rises in the TED spread and 3 month LIBOR – OIS spreads. In October, the CDS spread was six times the August level. The price of AIG's stock was 90% below its August level. The market lost confidence in AIG's ability to insure financial assets. The flight to safety is seen in columns 5 and 6. Treasury bill interest rates fell as the D-J industrial average fell and risk measures rose. The financial collapse was unexpected, sudden and pervasive.

This example leads one to ask the following questions. [Q1] What is an optimal liabilities/net worth or capital requirement for a large (non-life) insurer that balances the expected growth against risk? [Q2] What is an optimal risk? [Q3] What are theoretically founded early warning signals of a crisis? I explain why the application of stochastic optimal control (SOC)/dynamic risk management is an effective approach to answer these questions. The theoretically derived early warning signal of a crisis is the excess liability ratio, equal to the difference between the actual and optimal ratio. This SOC analysis should be used by those charged with surveillance of financial markets.

6.3 The Economics and Actuarial Literature

Since the Modigliani-Miller irrelevance proposition, and despite extensive research for decades, there is little consensus on how firms choose their capital structure (Shim 2010). Both the literature and regulations have departed greatly from the irrelevance proposition. The banking studies use the ratio of risk weighted assets/ total assets as a measure of portfolio risk (Aggarwal and Jacques 1998). It is based upon the view that portfolio risk is mostly determined by how assets are allocated and the different risk types. This is inadequate for an insurance industry, since the liability side of the portfolio must also be taken into account. Hence an alternative measure of portfolio risk for an insurer is the volatility of asset/liability ratio (Cummins and Sommer 1996). This measure takes into account the variance/ covariance matrix between the vectors of assets and liabilities. The appropriate level of capital is determined mostly by the insurer's risk exposure.

There is a variety of theories. The tradeoff theory postulates that firms seek to maintain a target/optimal debt ratio by balancing the tax advantages of debt financing against the costs of financial distress (Meyers 2001). The target ratios may vary from firm to firm. The pecking order theory of capital structure states that firms prefer retained earnings as the main source of financing new investment followed by debt financing followed by new issues of equity. Thus more profitable firms have less need for leverage. The financial theory suggests that capital structure is determined by firm specific characteristics such as profitability, growth opportunities and size of firm.

Korteweg (2010) reviews the literature on net benefits to leverage and then estimates the market's valuation of the net benefits to leverage by using panel data from 1994 to 2004, identified from market values and betas of the company's debt and equity. The literature tests the trade off between debt versus equity financing by running cross sectional regressions of leverage on a set of variables that proxy for the benefits and costs of leverage. The problem with this regression approach is that it is not possible to detect whether firms have too much or too little debt. The implicit assumption is that firms are on average optimally leveraged. This assumption lacks cogency in light of the analysis especially in Chap. 3. High profits mechanically lower leverage debt/equity so that cross section regressions show a negative relation between profitability and leverage, even though optimal debt ratios are positively related to profitability.

The article by Volchan (2007) reflects the state of the art in actuarial risk science. In Chap. 4, I focused upon Volchan's excellent discussion of the Cramér-Lundberg model, and I generalized the Classical gambler's ruin problems. My conclusions in Chap. 4 were that the Cramér-Lundberg analysis is inadequate to evaluate the risk/ return in the AIG case. First, the asset side of the equation for the change in surplus is ignored. The insurance company has assets against the liabilities that bring in income. Second, the value of the claims against AIG are highly negatively correlated with the value of the assets. When the market value of the insured assets decline, AIG must either compensate the insurer for the difference or put up more collateral. Third, the assets that are insured by AIG are quite closely correlated with the assets in AIG's portfolio.

6.4 Stochastic Optimal Control (SOC) Approach to Optimal Liabilities of a Large Insurer

The message from the actuarial literature above is that a risk measure should reflect the economic elements to measure risk, the economic situation, and that the problem is to find the optimal "required capital", liabilities/net worth, and risk to maximize expected surplus. Neither the economics literature nor the Cramér-Lundberg literature surveyed above answers questions [Q1]–[Q3] whereas the (SOC) approach is directly concerned with these questions. As described above, the AIGFP followed policies that were far from optimal, which led to the collapse of AIG. I derive a warning signal of financial difficulty by focusing upon the difference between the actual liabilities and the optimal liabilities.

SOC is a dynamic optimization, where key variables are stochastic, that I apply to the AIG case. Technical mathematical details are in the appendix to Chap. 4.

My criterion function is the maximization of the expected logarithm of surplus (net worth) at a future date. Surplus $X = A - L$ is assets A less L liabilities. This is: $V(X) = \max \ln X(T)$ at terminal date T. There are several advantages of this criterion. First, it is a risk averse strategy because the logarithm is a concave function. Unexpected bad events are weighted more heavily than unexpected good events. Second, the focus is upon the surplus X, which is an objective measure of economic performance. This measure of performance is relevant for stockholders, the parties that are insured by the firm and regulators. The control variable of the insurance company is the ratio of insurance liabilities/surplus. The ratio of assets/surplus = leverage is equal to one plus the ratio of liabilities/surplus.

The dynamics of the surplus, (6.1) is the sum of several components.

$$\begin{aligned} \textit{Change in surplus} &= (\textit{premium rate})\ \textit{liabilities} + (\textit{return} + \textit{capital gain})\ \textit{assets} \\ &\quad - (\textit{claims on the company related to liabilities}) \end{aligned} \tag{6.1}$$

The first term in the change in surplus is the premiums, equal to the premium rate times the insurance liabilities. The second term is the return on assets, which has two components. The first is the *productivity of assets*, which is deterministic, and the second is the *capital gain* on the assets. The third term concerns claims on the company that reduce the surplus. Claims may be losses to the insured parties and/or collateral calls by the parties that have insured assets whose Mark-to-Market (M-t-M) prices have fallen, as discussed above.

An increase in the liability ratio will increase expected growth of surplus if the return on investment plus premiums exceeds the claims. There are two stochastic variables in this model. First, the productivity of assets is observed, but the future capital gains are unknown when the investment decision is made. Second, the future claims are unknown when the insurance policies are sold. Third, there will be a correlation between the capital gain on assets and the claims. Most certainly, the correlation is negative. When asset prices, such as the index of housing prices fell, the claims against AIG increased, either for defaults or for collateral calls. The latter is what led to AIG's downfall.

The SOC approach derives an optimal liability ratio conditional upon the stochastic processes. The true stochastic process is unknown. Alternative stochastic processes imply different optimal liability ratios or optimal leverages. I start with a general model of the stochastic process and derive the optimal liability/surplus ratio, optimal risk, and optimal expected growth. Then I consider several specifications of the stochastic processes, as was done in Chaps. 4 and 5. These are special cases of the general model. In all cases, the capital gain and claims rate are negatively correlated. When asset prices fall, insurance claims rise.

A standard of optimality must be based upon sustainable stochastic processes. By contrast, the market optimized on the basis of unsustainable stochastic processes, which led to the bubble and its subsequent collapse.

Given the stochastic process, an optimal liability ratio is derived using the method explained in Chap. 4, making the appropriate changes. The expected growth of surplus or net worth is a concave function of the liability ratio, Fig. 4.2. It is maximal when the optimal ratio is chosen. As the liability ratio exceeds the derived optimal, the expected growth declines and the risk (variance) rises, Figs. 4.1 and 4.3. If the liability ratio is less than the optimal, expected growth is unduly sacrificed to reduce risk. Leverage is equal to one plus the liability ratio; and the capital requirement is the inverse of the leverage. I focus upon the liability ratio, and the optimal capital requirement and optimal leverage follow.

The main theoretical results are as follows. (1) The optimal liability ratio is not a number, but a function. It is proportional to: the drift of the capital gain less the drift of the claims plus the current productivity of capital plus the insurance premium rate less a risk premium. The factor of proportionality is the reciprocal of risk elements. Therefore the optimal liability ratio or capital requirement will vary over time. (2) Define the excess liability ratio as the actual liability ratio less the optimal ratio. For a sufficiently high excess liability or debt, the expected growth is zero or negative and the variance is high. See Figs. 4.1 and 4.2. *The probability of a decline in net worth or a debt crisis is directly related to the excess liability or debt ratio.* This is our Early Warning Signal (EWS) of a crisis.

6.5 Mathematical Analysis

The expected logarithm of the surplus of the insurer is the growth variable that is consistent with the insurance literature. Let $V(X,T)$ be the expected logarithm of surplus $X(T)$ at time T relative to its initial value $X(0)$. The stochastic optimal control problem is to select ratios $f(t) = L(t)/X(t)$ of liabilities $L(t)$ to surplus $X(t)$ during the period $(0,T)$ that will maximize $V(X,T)$ in (6.2). The maximum value is $V^*(X,T)$. Ratio $f^*(t)$ is the optimal leverage, or $1/f^*(t)$ is the optimal capital requirement and will vary over time. The stochastic optimal control tells us what is an optimal and what is an "excessive" leverage.

$$V^*(X,T) = max_f E\, In[X(T)/X(0)]$$
$$f = L/X = \text{liabilities/surplus},\ A/X = 1 + f = \text{leverage} \tag{6.2}$$

As explained in Chap. 4, an advantage of the criterion function equation (6.2) is that the model can be solved by using the Ito equation rather than dynamic programming, the role of the key variables in deriving the optimal capital requirement is clearly seen, it corresponds to an explicit economic objective and the model can be easily implemented for empirical purposes.

The change in surplus dX(t) in (6.3) is equal to the earnings of the insurer plus the revaluation of assets – the capital gains or losses – less the claims paid out. The *earnings* are: (1) the income from premiums $\pi L(t)$ equal to the premium π per dollar of insurance liabilities times L(t) the liabilities plus (2) the earnings from the assets $\beta A(t)$, where A(t) is the value of assets and β is the deterministic rate of return. (3) The *revaluation of assets,* the capital gain or loss on assets, is $[dP(t)/P(t)]A(t)$, where P(t) is the price index of the asset – such as the house price index in the AIG case. The change in price dP(t) and claims C(t) are stochastic. The first term in brackets is earnings and the second term in brackets is the capital gains. The length of the period is dt.

$$dX(t) = [\pi L(t)dt + \beta A(t)dt] + [(dP(t)/P(t))A(t)] - C(t). \tag{6.3}$$

The dynamics of the process to derive E[ln X(T)] is based upon (6.3). There are several possibilities for modeling the stochastic processes for the claims C(t) and capital gains dP(t)/P(t). I work with a general model, as I did in Chaps. 4 and 5, and then consider specifications as I did in Chap. 5.

Assume, for simplicity, that the premium rate π is given. At this premium rate there is an elastic demand for insurance and the insurer decides how much insurance L(t) to offer at that rate. L(t)/X(t) = f(t) *is the control variable*. Let the claims C(t) be described by stochastic differential equation (6.4). They are proportional to L(t) the amount of insurance liabilities. Claims are the required payments to the insured holders of CDS, *due to either defaults of the obligors or for collateral calls when the prices of the insured securities decline*. The latter led to the downfall of AIG. The mean of claims C(t) is cLtdt. The variance of the claims is $\sigma_c{}^2L^2(t)$ dt. Brownian Motion term dw_c has independent and stationary increments, with zero expectations, as assumed in the classical literature.

$$C(t) = [c\ dt + \sigma_c dw_c]\,L(t) \tag{6.4}$$

$$E(dw_c) = 0, E(dw_c)^2 = dt. \tag{6.4a}$$

Stochastic differential equation (6.5) concerns the capital gain/loss term dP(t)/P(t). The time varying drift is a(t)dt and the diffusion is $\sigma_p dw_p$. This is a general formulation. The variance of the capital gain is $\sigma^2{}_p$dt. The dw_p term has independent and stationary increments.

$$dP(t)/P(t) = a(t)dt + \sigma_p dw_p \tag{6.5}$$

$$E(dw_p) = 0, E(dw_p)^2 = dt. \tag{6.5a}$$

The two Brownian Motion terms are expected to be negatively correlated (6.6). Correlation coefficient $-1 \leq \rho < 0$. When there is a period with capital losses $dP(t)/P(t) < 0$, then claims against AIG including collateral calls are most likely to be high. This condition accurately described the period of the AIG crisis with CDS, Sect. 6.1.

$$E(dw_p\ dw_c) = \rho\ dt. \quad -1 \leq \rho < 0 \tag{6.6}$$

Using (6.4, 6.5 and 6.6) in (6.3), derive the stochastic differential equation (6.7) for the change in surplus dX(t). The first set in brackets contains the deterministic part and the second set contains the stochastic part: the randomness of the price of the assets less the randomness of the claims.

$$dX(t) = [\pi L(t) + \beta A(t) - cL(t) + A(t)a(t)]dt + [A(t)\sigma_p dw_p - L(t)\sigma_c dw_c]. \tag{6.7}$$

X = A − L = surplus, A = assets, L = liabilties, β = return on assets, π = premium rate, C = claims, a(t) = drift capital gain, stochastic BM terms dw_p, dw_c. Correlation between them is $\rho < 0$. Leverage = A/X = (1 + f)), debt ratio f = L/X.

I follow the analysis in Chaps. 4 and 5. The control variable is f(t) = L(t)/X(t) the ratio of liabilties/surplus. Write (6.7) as (6.8) in terms of the liability, or debt, ratio f(t), which is the control variable. Abbreviate it where terms M(f(t)) and N(f(t)) correspond respectively to the two terms in brackets in (6.8). Term M(f(t)) is deterministic and N(t) is stochastic with an expectation of N(f(t)) = 0.

$$\begin{aligned}
dX(t)/X(t) &= [\pi f(t) + \beta(1 + f(t)) - cf(t) + (1 + f(t))a(t)]dt + [(1 + f(t))\sigma_p dw_p \\
&\quad - f(t)\sigma_c dw_c] \\
dX(t)/X(t) &= M(f(t))dt + N(f(t)). \\
M[f(t)] &= [\pi f(t) + \beta(1 + f(t)) - cf(t) + (1 + f(t))a(t)] \\
N(f(t)) &= [(1 + f(t))\sigma_p dw_p - f(t)\sigma_c dw_c]. \\
E[N(f)] &= 0, E[M(f)dt, N(f)] = 0.
\end{aligned} \tag{6.8}$$

6.6 Solution for the Optimum Liability Ratio in General Model

As explained in Chap. 4/appendix, using the Ito equation, solve (6.8) for the expectation of d[ln X(t)]. This is (6.9) in terms of M(f(t)), which I call MEAN, and R(f(t)), which I call RISK. A great advantage of this approach is that it can be interpreted as a M-V "Mean-Variance" analysis, as was done in Figs. 4.1 and 4.2.

$$E[d\ ln\ X(t)] = M(f(t)) - (1/2)R(f(t)) \tag{6.9}$$

$$RISK\ R(f(t)) = \left[(1+f)^2\sigma_p{}^2 + f^2\sigma_c{}^2 - 2f(1+f)\rho\sigma_p\sigma_c\right] \quad (6.10)$$

$$MEAN\ M(f(t)) = [\pi f(t) + \beta(1+f(t)) - cf(t) + (1+f(t))a(t)] \quad (6.11)$$

The optimal liability/surplus f*(t) in (6.12) is the liability ratio that maximizes E[d ln X(t)] in (6.9). Term $\sigma^2 = (\sigma^2{}_p + \sigma^2{}_c - 2\rho\sigma_p\sigma_c)$ is in effect the variance of the difference between the capital gains and the claims, (dP/P − C(t)). The negative correlation $\rho < 0$ increases the risk.

The first term in brackets in (6.12) for the optimum ratio is $(\pi - c)$ the premium less the drift of the claims plus $(\beta + a(t))$ the sum of the deterministic rate of return on assets plus the time varying drift of the capital gain. The second term and the denominator involve the risks of the capital gains and claims.

$$\begin{aligned} f^*(t) &= argmax_f\{E[d\ ln\ X(t)] = [M(f(t)) - (1/2)R(f(t))]\} \\ &= [(\pi - c) + (\beta + a(t))] - (\sigma^2{}_p - \rho\sigma_p\sigma_c)]/(\sigma^2{}_p + \sigma^2{}_c - 2\rho\sigma_p\sigma_c) \quad (6.12) \\ \sigma^2 &= (\sigma^2{}_p + \sigma^2{}_c - 2\rho\sigma_p\sigma_c) \end{aligned}$$

6.7 Model Uncertainty and Optimal Liability Ratio

Section 6.1 explained that AIG sold CDS on reference securities/financial instruments that were primarily made up of subprime mortgages. AIG's downfall stemmed in a very large part from its CDS on multi-sector CDOs, which exposed the firm to the vaporization of the value in the subprime market. The value of the reference securities, the CDOs containing the subprime mortgages corresponds to A(t), and the value of the insurance liabilities CDS corresponds to L(t), in (6.3) and (6.4).

When the debt ratio f(t) exceeds f-max in Fig. 4.1, the expected growth is negative and the risk R(f) is high. Thus the excess debt f(t) − f*(t), the actual less the optimal, is a warning signal of a crisis. That is, leverage and risk are "too high". A crucial variable in evaluating the optimal debt ratio and risk is dP(t)/P(t) in (6.5). The question is how should one model and estimate the drift a(t) in that equation. This is the issue of model uncertainty.

Fisher Black (quoted in Derman), who would have won the Nobel prize for options modeling – had he lived – explained that a financial theory: "is accepted not because it is confirmed by conventional empirical tests, but because researchers persuade one another that the theory is correct and relevant." Black argues that given the models' limitations, "the right way to engage with a model is, like a fiction reader or a really great pretender, is to suspend disbelief and push it as far as possible... But then, when you've done modeling, you must remind yourself that ... although God's world can be divined by principles, humanity prefers to remain mysterious. Catastrophes strike when people allow theories to take on a life of their own and hubris evolves into idolatry."

Fig. 6.1 HPI/CAPGAIN Capital gains [P(t) − P(t − 1)]/P(t − 1), percent change in index of house price HPI, sample 1991q1–2011q1

The question is what is the optimal liability and the excessive liability that should have been a warning signal for the crisis. How should we model $dP(t)/P(t) = a(t)\,dt + \sigma_p dw_p$ that corresponds to HPI/CAPGAIN $= [P(t+1) - P(t)]/P(t)$? Is the capital gain equal to a constant drift $a(t) = a$ plus a Brownian Motion term, or is the drift ergodic mean reverting to a constant trend plus a Brownian Motion term? Are the capital gains terms independent or are they serially correlated? How can we characterize the distribution of the capital gains?

I approach the problem of deciding on the appropriate way to model the capital gain equation by looking at Fig. 6.1. It plots the appreciation of single-family housing prices, CAPGAIN/HPI, a four quarter appreciation of US Housing prices HPI, percent p.a. It corresponds to $dP(t)/P(t)$. Figure 6.1 covers a recent period 1991q1–2011q1, pre and post, crisis. Consider, however, the information available *before* the crisis. What can one say about the distribution of the change in house prices over the period 1980q1–2007q4?

From Figs. 3.4 and 5.2, we can reject the hypothesis that the distribution of the CAPGAIN series is normal. Table 6.2 tests for serial correlation of the capital gains. Are they independent? The Q-Statistics reject the null hypothesis that there is no autocorrelation up to order k. This means that the Brownian Motion assumption used in modeling dP/P is questionable.

In Chaps. 4 and 5, two models for the capital gain term were considered. In model I the price $P(t)$ was ergodic mean reverting to a constant trend plus a BM term. In model II, the capital gains term was BM with a constant drift. The choice between the models depends upon a unit root test. Table 6.3, which tests for a unit root, is concerned with the question whether or not the capital gains term is stationary. This is the null hypothesis.

Table 6.2 Tests for autocorrelation

CAPGAIN(t) against CAPGAIN (t-k)
Sample: 1980Q1 2007Q4
Included observations: 111

		Lag	AC	PAC	Q-Stat	Prob
Autocorrelation AC at lag k = Sum [covariance Y(t + k), Y(t)]/variance Y(t)	Q-Stat probability = probability of null hypothesis that there is no autocorrelation up to order k	1	0.929	0.929	98.423	0.000
		2	0.847	−0.118	180.98	0.000
		3	0.748	−0.164	245.97	0.000
		4	0.628	−0.204	292.17	0.000
		5	0.545	0.246	327.30	0.000
		6	0.466	−0.028	353.20	0.000
		7	0.390	−0.087	371.51	0.000
		8	0.331	−0.011	384.89	0.000

Table 6.3 Unit root tests

Null hypothesis: CAPGAIN has a unit root;
Ho : r = 1 CAPGAIN(t) = r CAPGAIN(t − 1) + e(t).
Exogenous: Constant
Lag length: 5 (automatic based on SIC, MAXLAG = 8)

		t-Statistic	Prob.[a]
Augmented Dickey-Fuller test statistic		−2.662054	0.0841
Test critical values	1% level	−3.493747	
	5% level	−2.889200	
	10% level	−2.581596	

[a]MacKinnon (1996) one-sided p-values

$$\text{Ho}: \text{r} = 1, \ \text{CAPGAIN(t)} = \text{r CAPGAIN(t} - 1) + \text{e(t)}, \text{e(t) random.}$$

If the null r = 1 is not rejected one can conclude that a BM with drift is the appropriate way to model the capital gains equation. If the null hypothesis is rejected, the appropriate way to model the capital gains equation is ergodic mean reversion with a BM term. The conclusion of the unit root test in Table 6.3 is that: One cannot reject the null hypothesis at the 5% level, but can at the 10% level. *Neither hypothesis is inconsistent with the data*. Since there is model uncertainty, my approach is to consider a general formulation for the optimal debt liabilities in (6.12) and compare it with the actual debt liabilities.

6.8 Early Warning Signals

The optimal debt ratio $f^*(t)$ is positively related to the drift of the *capital gain* and negatively related to the drift of *claims*. The negative correlation between the capital gains and the claims reduces the optimal debt ratio, because it increases the total risk.

Fig. 6.2 $\Psi(t) =$ EXCESSDEBT, normalized excess debt ratio of the mortgagors/households

The early warning signal follows the analysis in Chaps. 4 and 5. The excess debt of the mortgagors/households is the difference between the actual debt ratio and an upper bound $f^{**}(t)$ of the optimal liability ratio. The use of $f^{**}(t)$ the upper, bound allows the analysis to be done on the assumption that either model is correct. $\Psi(t) =$ EXCESSDEBT $= [f(t) - f^{**}(t)]$ is plotted below as Fig. 6.2. The greater is the value of $\Psi(t)$: (a) the lower is the expected growth of surplus and the greater is the risk, as is seen in Fig. 4.1 and (b) it is more likely is it that households/mortgagors will default on the mortgages thereby increasing the claims on AIG.

My strategy is to compare this measure $\Psi(t) =$ EXCESSDEBT of the mortgagors with the actual liabilities $L(t)$ of AIG to derive a warning signal. *If actual liabilities* $L(t)$ *are rising* when $\Psi(t)$ *is increasing, the liklihood of claims on AIG and a crisis is more probable.*

Consider AIG liabilities $L(t)$ or $f(t)$. AIG's business of offering credit protection on assets such as mortgage backed securities and CDO's grew from $20 billion in 2002 to $211 billion in 2005 and $533 billion in 2007. (FCIC, p. 141). The company as a whole had a much higher leverage ratio than did other insurance companies. As of December 2007, leverage for AIG was 11:1, whereas leverage for Berkshire Hathaway 2:1, Travelers 4:1 and Chubb 4:1. A 9% decline in the value of assets of AIG would wipe out surplus.

The EXCESSDEBT rose considerably from 2002 to 2007 by more than two standard deviations, whereas the actual AIG liabilities rose, $L(2007)/L(2002) = 533/20 = 26.6$ during this period. This means that the probability of large claims was very high. When the value of house prices $P(t)$ declined, the value of the CDOs insured by AIG declined. AIG's credit rating was downgraded, it had to post additional collateral or compensate the insured for the difference between the notional and market values of the reference securities. However, AIG lacked the liquidity to meet the collateral calls and banks

Table 6.4 Notional amount CDS outstanding $billions, AIG- CDS rate 2001 h1–2010 h2

Date, half year	(1) Notional amount CDS outstanding, $billions (rounded)	(2) AIG-CDS rate
2001 h1	631	15.57
2001 h2	918	22.64
2002 h1	1,563	37.3
2002 h1	2,191	64.11
2003 h1	2,687	40.4
2003 h2	3,779	27.74
2004 h1	5,441	22.11
2004 h2	8,422	21.25
2005 h1	12,429	30.7
2005 h2	17,096	23.8
2006 h1	26,005	18.13
2006 h2	34,422	11.82
2007 h1	4,564	11.5
2007 h2	62,173	48.95
2008 h1	54,611	153.7
2008 h2	38,563	938.8
2009 h1	31,223	1,487.9
2009 h2	30,428	935.6
2010 h1	26,263	417.4

refused to refinance. This led to the crisis of AIG. Section 6.1 and Table 6.1 show that the crisis was not predicted and came as a shock-contrary to Gorton's model (see part 1 above). On the other hand, the SOC analysis provided a good early warning signal.

The market did not exhibit qualms about AIG. Table 6.4 shows the total notional amounts of CDS outstanding in column (1). Column (2) is the CDS rate on AIG, which reflects the probability that AIG will default. Note the AIG-CDS rate declined from 30.7 in 2005 h1 to 11.5 in 2007 h1, while the notional total of CDS outstanding rose from $12429 billion in 2005 h1 to $62173 billion in 2007 h2. The price of AIG insurance declined while (a) the total amount of insurance increased and (b) The EXCESSDEBT rose considerably from 2002 to 2007 by more than two standard deviations. The market was taken by surprise because it ignored available information on household debt.

6.9 An Evaluation of the Bailout

The COP Report (2010) evaluates the Federal Reserve and Treasury justifications for the bailout. I paraphrase this Report, which is a suitable closing for this chapter.

6.9.1 Government's Justification for Rescue

The Federal Reserve was the only governmental entity at the time with the legal authority to provide liquidity to the financial system in emergency circumstances. The Board and FRBNY ultimately chose to provide AIG with assistance after identifying the systemic risks associated with the company and contemplating the consequences of an AIG bankruptcy or partial rescue. On September 16, the Board, with the full support of Treasury, authorized Federal Reserve Bank of New York (FRBNY) to lend up to $85 billion to AIG in order to assist the company in meeting its obligations as they came due. The Board determined that, in the then-existing environment, "a disorderly failure of AIG could add to already significant levels of financial market fragility and lead to substantially higher borrowing costs, reduced household wealth, and materially weaker economic performance."

The Federal Reserve's actions were based upon its judgment that an AIG collapse would have been much more severe than that of Lehman Brothers because of its global operations, substantial and varied retail and institutional customer base, and the various types of financial services it provided. At September 2008 the time of the initial decision to assist AIG, the Federal Reserve and Treasury publicly identified three primary ways in which an AIG failure posed systemic risk. First, they concluded that, given AIG's role as a large seller of CDSs on CDOs, an AIG failure could have exposed its counterparties to large losses and disrupted the operation of the payments and settlements system. According to Secretary Geithner, if the AIG parent holding company had filed for bankruptcy, defaults on over $100 billion of debt and on trillions of dollars of derivatives would have resulted. The Federal Reserve and Treasury argued that this would have adversely impacted numerous financial institutions and the financial system as a whole. The primary fear of the Federal Reserve and Treasury was that defaults directly related to AIG would have spread throughout the financial system, affecting transactions between other counterparties, negatively affecting investor confidence, and further destabilizing the economy. Second, the Federal Reserve and Treasury concluded that an AIG default on its commercial paper could have adversely impacted money market mutual funds since AIG had issued $20 billion in commercial paper to money market mutual funds, approximately four times as much as Lehman Brothers. In the government's view, this could have substantially disrupted the commercial paper market by reducing credit availability for borrowers even on a short-term basis and causing higher lending rates. This concern escalated after the money market disruptions that occurred in the wake of the Lehman Brothers bankruptcy filing, including the "breaking of the buck" at the Reserve Primary Fund.

Third, the Federal Reserve and Treasury asserted that they feared that an AIG failure could have undermined an already fragile economy by weakening business and investor confidence. After the placement of Fannie Mae and Freddie Mac into government conservatorship on September 7 and the Lehman Brothers bankruptcy filing on September 15, financial markets destabilized considerably. AIG maintained financial relationships with a large number of banks, insurance companies, and other market participants across the globe. A failure of AIG in this environment, according to the Federal Reserve and Treasury, could have further shaken investor confidence and contributed to increased borrowing costs and additional economic deterioration.

In this context, the Federal Reserve and Treasury officials stated that they believed that the unfolding crisis and the increasingly fragile state of the economy necessitated swift action to prevent a total collapse of the financial system.

6.9.2 Panel's Analysis of Options Available to the Government and Decisions

The Panel recognized that policymakers faced a deepening financial crisis and that there were many issues of serious concern and a limited amount of time in which to respond to AIG in September 2008. The Panel stated that FRBNY's decisions were made in the belief that it alone could act and that it had to choose between options that were all unattractive. There is nothing unusual about central banks acting as the lender of last resort. However, by adopting the terms developed by the private sector consortium and retaining most of its terms and conditions, FRBNY chose to act in effect as if it were a private investor in many ways, when its actions also had serious public consequences whose full extent it may not have appreciated. FRBNY also failed to recognize the AIG problem and get involved at a time when it could have had more options.

While the reasons for FRBNY's failure are not clear, it is clear that when FRBNY finally realized AIG was failing and that there would be no private sector solution, *Chairman Bernanke and President Geithner failed to consider any options other than a full rescue. To have the government step in with a full rescue was not the approach used in prior crises, including Bear Stearns and Long-Term Capital Management.* FRBNY chose lawyers from a limited pool and did not seek legal advice from a debtor's counsel (such as AIG's bankruptcy counsel or independent bankruptcy counsel). As a result, there were many options FRBNY evidently did not consider, including a combined private/public rescue (which would have maintained some market discipline), a loan conditioned on counterparties granting concessions, and a short-term bridge loan from FRBNY to provide AIG time for longer-term restructuring. *Providing a full government rescue with no shared sacrifice among the creditors who dealt with AIG fundamentally changed the relationship between the government and the markets, reinforcing moral hazard and undermining the basic tenets of capitalism.* The rescue of AIG dramatically added to the public's sense of a double standard – where some businesses and their creditors suffer the consequences of failure and other, larger, better connected businesses do not. The FRBNY's decision-making also suggest that it neglected to give sufficient attention to the crucial need – more important in a time of crisis than ever – for accountability and transparency.

The Panel asked: Was It Truly an All-or-Nothing Choice? The Treasury/Fed presented the decision to rescue AIG as an all-or-nothing "binary" decision. In other words, the government asserted that it was necessary to rescue AIG in its entirety or let it fail in its entirety. It was not possible to pick and choose which businesses or subsidiaries could be saved. The Panel tested this assertion and considered whether bankruptcy had to be an all-or-nothing option, in terms of the entities covered, the obligations covered, or in terms of timing. If a bankruptcy was not a real option in September 2008, was it later?

The potential impact of an AIG bankruptcy can be guessed by examining how the markets continued to deteriorate even after AIG was rescued. As shown in Table 6.1, the spread between the London Interbank Offered Rate (LIBOR) and the Overnight Index Spread Rate (OIS) – used as a proxy for fears of bank bankruptcy – dramatically increased in September 2008 amid the growing concerns of financial collapse. Former Federal Reserve Chairman Alan Greenspan stated that the "LIBOR-OIS spread remains a barometer of fears of bank insolvency." In the immediate aftermath of the Lehman bankruptcy this spread spiked to a level indicating actual illiquidity in the interbank market – not merely a high cost for obtaining funds – meaning that banks were not willing to lend to one another. Prior to the beginning of the credit market crisis in August 2007, the LIBOR-OIS spread was 10 basis points. The LIBOR-OIS spread reflected the contraction of liquidity that crippled the financial markets in 2008, as seen in Table 6.1.

The Panel recognized that the government was faced with a deepening financial crisis, and its attention was on a number of troubled institutions besides AIG in the course of just a few days. Given this context, the government took actions that it thought would facilitate rapid intervention in the midst of deteriorating economic conditions. Nonetheless, if the government concluded that it could not impose conditions on its assistance once it had decided to backstop AIG with taxpayer funds, or that other possible rescue alternatives were unattractive or impracticable, then it had an obligation to fully explain why it decided what it did, and especially why it was of the opinion that *all AIG's creditors and counterparties would receive all amounts they were owed.* In addition, while the Panel acknowledged the number of complex issues and troubled institutions that policymakers were concerned with at the time, it appears that the government was neither focused on nor prepared to deal with the AIG situation. By placing a tremendous amount of faith in the assumption that a private sector solution would succeed in resolving AIG, the government had no legitimate alternative on the table once that assumption turned out to be incorrect. In its assessment of government actions to deal with the current financial crisis, the Panel has regularly called for transparency, accountability, and clarity of goals. These obligations on the part of the government do not vanish in the midst of a financial crisis. In fact, it is during times of crisis, when difficult decisions must be made, that a full accounting of the government's actions is especially important.

6.10 Conclusions: Lessons to be Learned

From 1994 to 2006 AIGFP operating income grew by 18% pa paralleling the boom in the overall derivative market as well as in the CDS market. The risk involved in this business was dramatically disproportionate to the revenue produced. Losses in *one year* 2007 totaling $11.5 billion were twice the aggregate net income produced by the division over the *entire period* 1994–2006.

AIGFP relied on Gorton's actuarial model that did not provide a tool for monitoring the CDO's market value. Gorton's model had determined with 99.85% confidence that the owners of super-senior tranches of the CDO's insured by AIGFP would never suffer real economic losses. The company's auditor PWC apparently was also not aware of the collateral requirements. PWC concluded that "..the risk of default on the [AIG] portfolio has been effectively removed as a result of a risk management perspective...". AIG was led to accumulate staggering amounts of risk, especially in its Financial Products subsidiary.

What lessons can be learned from AIGs experience? AIG made several serious mistakes. First, the financial structure of derivatives CDS insured by AIG, was ultimately based upon the systemic risk from the inability of the mortgagors to service their debts. The exaggerated view of the drift of the capital gain, which drove the bubble, was based upon the growth of the value of housing, the housing price index 2004–2006. This growth ranging from 10% to 14% pa exceeded the mean interest rate. Mortgagors consumed the difference between the refinancing of the loan and the associated debt. They expected to service their debts from the appreciation of the housing price in excess of the rate of interest. *This free lunch would disappear when the unsustainable capital gain declined below the interest rate.*

Second, risk was underestimated because AIG ignored the negative correlation between the capital gain and the claims. The CDS claims grew when the value of the insured obligations CDO declined. This set off collateral requirements and the stability of AIG was undermined.

Acting Director of the OTS Polakoff acknowledged that AIG and OTS did not foresee the extent of risk concentration and profound systemic impact CDS caused within AIG. The AIG crisis leads one to ask the following questions. What is an optimal liability ratio or capital requirement for a large insurer of financial derivatives that balances the expected growth against risk? What is an optimal risk? What are theoretically founded early warning signals of a crisis? Neither the economics literature nor the Cramér-Lundberg literature answers these questions. I explained why the application of stochastic optimal control (SOC)/dynamic risk management is an effective approach to answer these questions. The theoretically derived early warning signal of a crisis is the excess liability ratio, equal to the difference between the actual and optimal ratio. This SOC analysis should have been used by those charged with surveillance of financial markets.

References

Aggarwal, R., and K.T. Jacques. 1998. *Assessing the impact of prompt action on bank capital and risk*. New York: Federal Reserve Bank of New York, Economic Policy Review.
Congressional Oversight Panel. 2010. *The AIG rescue, its impact on markets and the government's exit strategy*. Washington, DC: U.S.G.P.O.
Cummins, J.D., and D.W. Sommer. 1996. Capital and risk in property–liability insurance markets. *Journal of Banking and Finance* 20: 1069–1092.
Derman, Emanuel. 2004. *My life as a quant*. Hoboken: Wiley.
Financial Crisis Inquiry Report, FCIC Final Report of National Commission. 2011. Public Affairs, New York.
Korteweg, Arthur. 2010. The net benefits to leverage. *Journal of Finance* LXV(6): 2089–2136.
MacKinnon. 1996. *Journal of Applied Econometrics*.
Meyers, S.C. 2001. Capital structure. *Journal of Economic Perspectives* 15: 81–102.
Meyers, S.C., and J.A. Read. 2001. Capital allocation for insurance companies. *Journal of Risk and Insurance* 68: 545–580.
Shim, Jeungbo. 2010. Capital-based regulation, portfolio risk and capital determinants: Empirical evidence from US property insurers. *Journal of Banking and Finance* 34: 250–261.
Stein, Jerome L. 2010. Greenspan's retrospective of financial crisis and stochastic optimal control. *European Financial Management* 16: 5858–5871.
Volchan, S. 2007. The fundamental theorem of actuarial risk science. *The Mathematical Scientist* 31(2): 73–84.

Chapter 7
Crisis of the 1980s

Abstract There have been three crises since the 1930s: the Agricultural crisis of the 1980s, the S&L crisis in the 1980s and the major financial crisis of 2008. The two crises of the 1980s had many features in common, but were localized. The crisis of 2008 shared the common elements of the earlier two but was more pervasive and severe due to the financial structure that was based upon the housing/mortgage sector. I focus upon the agriculture crisis. How should creditors, banks and bank regulators evaluate and monitor risk of an excessive debt that significantly increases the probability of default? I show how the techniques of stochastic optimal control used in earlier chapters provided early warning signals of the crisis. I compare the S&L crisis to the agricultural crisis.

7.1 Introduction

There were three crises since the 1930s: the Agricultural crisis of the 1980s, the S&L crisis in the 1980s and the major financial crisis of 2008. The two crises of the 1980s had many features in common, but were localized. The crisis of 2008 shared the common elements of the earlier two but was more pervasive and severe due to the financial structure that was based upon the housing/mortgage sector. In the current chapter I focus upon the crises of the 1980s, in particular the agriculture crisis. This chapter addresses the question: *How should creditors, banks and bank regulators evaluate and monitor risk of an excessive debt that significantly increases the probability of default?* I show how the techniques of stochastic optimal control used in earlier chapters are useful in providing early warning signals. In the concluding part I compare the S&L crisis to the agricultural crisis.

The agricultural debt crisis of the 1980s is emblematic of the bubble-crisis phenomenon. Figure 7.1 graphs the ratio of debt/equity DEBTEQUITY and the delinquency rate on loans DELIQRATEFCS. The variables have been normalized to facilitate comparison. N(Z) is (Z-mean)/standard deviation. The agricultural debt crisis is the shaded area in Fig. 7.1.

J.L. Stein, *Stochastic Optimal Control and the U.S. Financial Debt Crisis*,

DOI 10.1007/978-1-4614-3079-7_7, © Springer Science+Business Media New York 2012

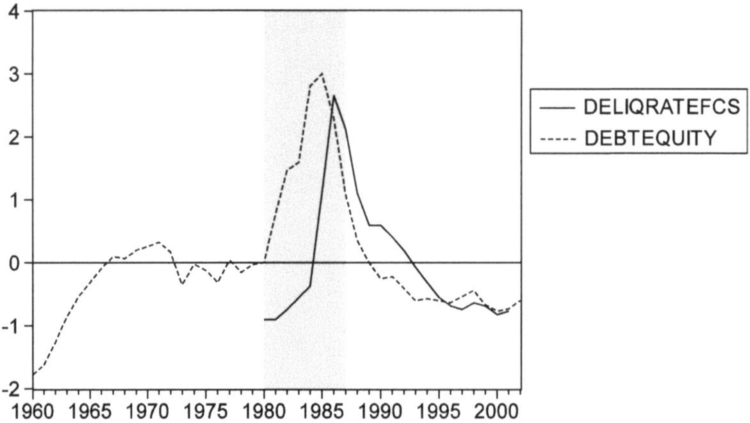

Fig. 7.1 Ratio debt/equity (DEBTEQUITY), delinquency rate (DELIQRATEFCS) of the agricultural sector. Normalized variables. A normalized variable N(Z) is (Z-mean)/standard deviation. Thus the mean is zero and the standard deviation is unity (*Source*: Farm Security Administration, Economic Research Service USDA)

Table 7.1 Histogram of debt/equity ratio (Source of data: Economic Research Service USDA, Agricultural Income and Finance Outlook/AIS-77/February 26, 2002)

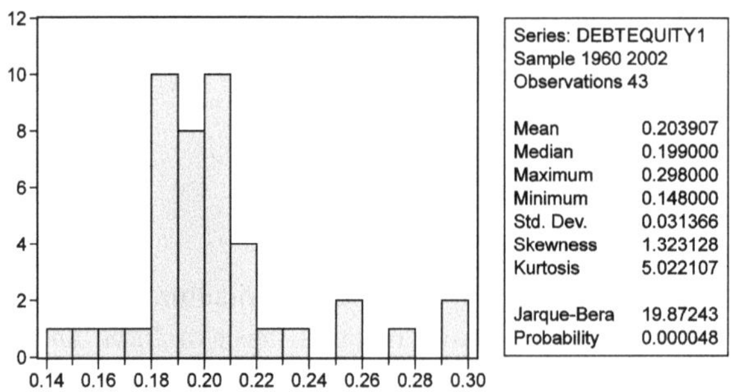

The debt ratio in the agricultural sector is stable until the 1980s and then spikes at 1985 at more than 2.5 standard deviations above its mean. This corresponds to the bubble, which bursts in the second half of the decade. The bursting is reflected in the delinquency rate on loans. The debt ratio returns in 1990 to its mean, the longer term level. The delinquency rate on loans follows a similar pattern.

Table 7.1 is a histogram of the debt ratio in the agricultural sector. It is not normally distributed, with outliers in the boom period. The low probability value of the Jarque-Bera statistic leads to a rejection of the null hypothesis of a normal distribution.

The focus in this chapter is upon the banks that financed the asset purchases by the agricultural sector. Banks and bank regulators, such as the FDIC, should evaluate whether the borrowers are likely to default. *When is the debt ratio "excessive" so that delinquency is likely?* What were theoretically based, objectively measured, early warning signals that should have signaled the agricultural crisis and delinquencies? I apply the same techniques of SOC used in the previous chapters, to derive the optimal debt in the agricultural sector where there are risks on both the asset and liability sides of the balance sheet. The ratio of debt/net worth per se is not a significant explanation of defaults. The vulnerability of the system to shocks, from either the return to capital, the interest rate or capital gain, increases directly with the excess debt, which is defined as the actual less the optimal debt ratio. As the excess debt rises, risk rises relative to expected return and default becomes more likely. In this manner the SOC approach developed here should be useful to the FDIC and other bank regulators.

I draw upon the study by the FDIC (1997) to discuss the history of the period: the cause and development of the bubble and the reason for its collapse.[1] Agriculture flourished in the 1980s. Farm exports grew rapidly and along with the domestic inflation farm incomes reached all-time highs. These factors produced capital gains on farm assets. Equity rose significantly. Credit was readily available. Real interest rates were low and farmers used the rising value of farm assets as collateral for loans. Farmers would purchase farm real estate with moderate down payments and, after the value of the newly purchased land increased, would use the increased equity to buy additional farmland with minimal down payments. Higher levels of real estate debt were supplemented by debt to finance machinery and equipment.

The sequence is described by Fig. 7.2. The speculation in land produced capital gains and raised the market value of equity (EQUITY). Consequently, the ratio of interest service on the debt/value added (INTVA), the debt burden, rose significantly. In the fall of 1979, the Federal Reserve undertook a restrictive monetary policy in order to reduce inflation, and interest rates rose drastically. The resulting appreciation of the US dollar reduced foreign demand for US agricultural products. The decline in foreign demand was exacerbated by the debt crisis in the less developed countries. Farm exports declined by 40% from 1971 to 1976, at a time when productive capacity had increased. The result was an accumulation of huge surpluses of farm commodities in the early 1980s. When the bubble collapsed in 1985, asset values and equity fell drastically. The resulting rise in the debt burden was devastating, and the (DELINQRATEFCS) delinquency rate on loans rose drastically.

Table 7.2 presents Granger causality tests. Did the return on assets, the sum of the capital gain and the productivity of capital, "cause" the debt ratio to rise? Did the rise in the debt ratio "cause" higher capital gains and raise the return? It seems that at best the rise in the return – capital gain – Granger caused a rise in the debt/equity ratio. This result is consistent with the story described above and graphed in Fig. 7.2.

[1] See Barry and Robison (2001) and Moss et al. (2003) for a discussion of the agriculture economics literature.

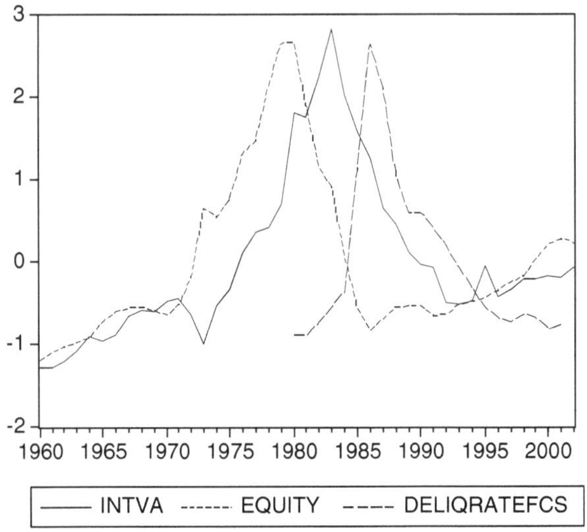

Fig. 7.2 Agricultural bubble and collapse. Normalized variables. INTVA = interest payments/value added = debt burden. DELIQRATEFCS = delinquency rate, farms security administration, as a percent of loans. EQUITY = assets − liabilities (*Source*: USDA (2002), Economic Research Service, Agriculture Income and Finance, Farm Income and Balance Sheet Indicators. Normalized variables)

Table 7.2 Granger causality tests. Return = capital gain plus productivity of capital, against the debt/equity ratio

Pairwise Granger causality tests
Time: 10:13
Sample: 1960–2002
Lags: 2

Null hypothesis	Obs	F-statistic	Probability
DEBTEQUITY does not Granger cause RETURN	41	1.14265	0.33027
RETURN does not Granger cause DEBTEQUITY		2.75373	0.07711

The agricultural crisis was localized, unlike the US financial debt crisis discussed in Chap. 5. The reason was that a structure of derivatives was not based upon the value of farmland, and there was no shadow banking system that was vulnerable to declines in farm asset values.

7.2 Stochastic Optimal Control (SOC) Analysis

I use the same SOC analysis presented in Chaps. 4, 5 and 6 to derive early warning signals of the agricultural debt crisis. The question is how should bank regulators view the debt ratio. *When is the debt ratio "excessive" so that delinquency is likely?*

The components of the analysis are the criterion function and the stochastic differential equations.

7.2.1 The Criterion Function

The lender or bank regulator should evaluate what debt ratio maximizes the expected (E) growth rate of the borrower's net worth over the period of the loan, an horizon of length T from the present t = 0. This would be the optimal debt that a prudent lender would want to offer. The bank/lender wants to avoid borrower's bankruptcy (X = 0) by placing a very high penalty on a debt that would lead to a zero net worth, bankruptcy. The lender is very risk averse.

The borrower has a net worth X(t) equal to the value of assets (used interchangeably for capital) K(t) less debt L(t). Initially net worth X = X(0) > 0. Equation 7.1 is the criterion function. The maximization is over the debt ratios f = L/X.

$$W(X,T) = max_f E\, ln[X(T)/X(0)], \quad X = K - L > 0, f = L/X. \tag{7.1}$$

The next steps are to: explain the stochastic differential equation for net worth, relate it to the debt ratio, and specify what are the sources and characteristics of the risk and uncertainty.

7.3 Dynamics of Net Worth

In view of (7.1), the bank/lender should focus upon the change in net worth dX(t) of the borrower. It is the equal to the change in capital dK(t) less the change in debt dL(t). Capital K = PQ, the product a physical quantity Q times the relative price P of the capital asset to the price of output. The change in the value of capital has two components. The first is the change due to the change in relative price of capital, which is the capital gain or loss, K(dP/P) term. The second is investment, which is I = P dQ, the change in the quantity times the relative price. The change in debt dL is the sum of expenditures less income. Expenditures are the debt service r(t)L(t) at real interest rate r(t), plus investment I = P dQ plus either consumption, dividends or distributed profits C(t). Income Y(t) = β(t)K(t) is the product of capital times β(t) its productivity. Variable b(t) is defined in (7.2.1) as dP(t)/P(t) + β(t).

For simplicity, assume that consumption C(t) is a constant fraction c > 0 of net worth X(t), *C(t) = cX(t)*. Combining these effects, the change in net worth is (7.2).

Change in net worth = value of assets (capital gain + productivity of capital) – (interest payments – consumption). Assets = K(t) P(t)Q(t). Interest payments r(t)L(t), debt L(t), r(t) interest rate.

$$dX(t) = K(t)[(dP(t)/P(t)) + \beta(t)] - r(t)L(t)dt - C(t)dt \tag{7.2}$$

$$b(t) = (dP(t)/P(t)) + \beta(t) \tag{7.2.1}$$

Stochastic variables are the real capital gain or loss (dP/P), the productivity of capital $\beta(t)$ and r(t) the real interest rate. Term b(t) in (7.2.1) subsumes the two sources of risk on capital: the capital gain or loss and the productivity of capital. The agricultural debt crisis can be understood in terms of (7.1) and (7.2).

7.4 The Stochastic Processes

A crucial assumption motivating the use of SOC is that *the future is unpredictable*. The uncertainty may have different forms and concerns the terms in (7.2). The object is to select a time varying debt ratio f(t) = L(t)/X(t) that will maximize (7.1) subject to the stochastic differential equation (7.2). The question is how to model the stochastic processes?

The uncertainty may have different forms. Since there is some ambiguity about describing the specific form of the stochastic processes underlying the variables in (7.2) whose statistics are in Table 7.3 and graphed in Figs. 7.3 and 7.4, I consider several cases summarized in Box 7.1.

Case A assumes that the return on capital b(t) = (dP/P) + $\beta(t)$ follows an Ornstein-Uhlenbeck process, Erogodic Mean Reversion (EMR) (7.3/A). The solution of this equation implies that the return b(t) converges to a distribution with a mean of b, and a variance of $\sigma_b^2/2\alpha$, where α is the speed of response. The interest rate r(t) is (7.4/A), a Brownian Motion with Drift (BMD) process. The mean is r dt and the variance is σ_r^2 dt. *Case B* assumes that both the return to capital (Eq. 7.3/B), and the interest rate (Eq. 7.4/B) are described by Brownian Motion with Drift.

Table 7.3 Basic statistics: debt/equity ratio DEBTEQUITY = L(t)/X(t), gross value added/capital $\beta(t)$= Y(t)/K(t) = GVACP, INTDEBT = interest expenses/debt = r(t) and RETVAINTD1 = GVACAP − INTDEBT = Y(t)/K(t) − r(t)

	DEBTEQUITY	GVACAP	INTDEBT	RETVAINTD1
Mean	0.2039	0.1041	0.0801	0.0240
Median	0.1990	0.1075	0.0771	0.0211
Maximum	0.2980	0.1372	0.1157	0.0681
Minimum	0.1480	0.0693	0.0584	−0.0350
Std. Dev.	0.03136	0.0144	0.0169	0.0258
Skewness	1.3238	−0.4462	0.5543	−0.4083
Kurtosis	5.0221	2.7084	2.1654	2.5861
Jarque-Bera	19.872	1.5791	3.4515	1.5017
Probability	0.000048	0.454044	0.178033	0.471962
Sum	8.768	4.480	3.447	1.032
Sum Sq. Dev.	0.0413	0.0087	0.0120	0.0281
Observations	43	43	43	43

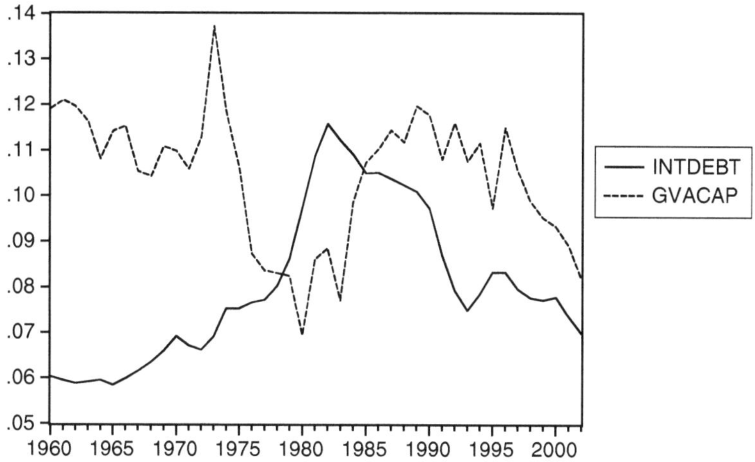

Fig. 7.3 Agriculture. GVACAP = gross value added/capital = productivity of capital = β(t), INTDEBT = total interest payments/debt = r(t)

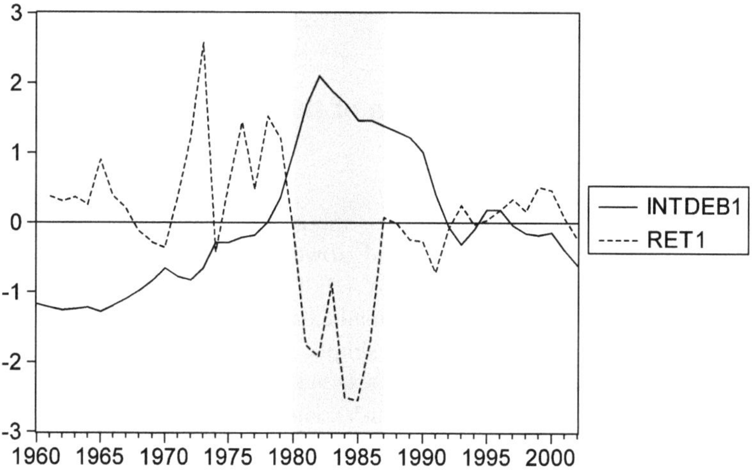

Fig. 7.4 RET1 = GVACAP + GROWASSETS = b(t), INTDEB1 = r(t)

Box 7.1 Stochastic Processes, Uncertainty

b(t) = dP(t)/P(t) + β(t)

Case A
Ornstein-Uhlenbeck (EMR)

$$db(t) = \alpha(b - b(t))dt + \sigma_b dw_b \quad \infty, > \alpha > 0 \qquad (7.3/A)$$

$lim\ b(t) \sim N(b, \sigma^2{}_b/2\alpha)$

(continued)

Brownian Motion-Drift (BMD)

$$r(t) = r\ dt + \sigma_r\ dw_r \quad (7.4/A)$$

Case B
Brownian Motion-Drift (BMD)

$$b(t) = b\ dt + \sigma_b\ dw_b \quad (7.3/B)$$

Brownian Motion-Drift (BMD)

$$r(t) = r\ dt + \sigma_r\ dw_r \quad (7.4/B)$$

Case C is the reverse of Case A
The mean (b, r) is written without any time index

7.5 Solution and Interpretation of the Optimal Debt/Net Worth

The solution for the optimal debt ratio – the control variable f(t) = L(t)/X(t) – concerns the maximization of the expected growth rate of net worth, (7.1) subject to the appropriate stochastic process in Box 7.1. There are other reasonable criteria functions and stochastic processes. The mathematical techniques for their solution are discussed in the Chap. 4. Here, I simply state the results. The basic data underlying the variables in Box 7.2 are discussed in Sect. 7.6. In Sect. 7.7, I explain the results stated in Box 7.2 using a Mean-Variance analysis. Then I use the results in discussing the agricultural debt crisis.

The optimal debt/net worth ratio f*(t) = L(t)/X(t) in Box 7.2 varies according to the three cases (A), (B) and (C) respectively. The asterisk denotes the optimal value. In each case, the numerator is a return less an interest rate, and the denominator is a variance.

In *case (A)*, (7.5/A), the optimal ratio of debt/net worth f*(t) varies with time. It is equal to the *current* value of the return to capital b(t) less the *mean* rate of interest r, divided by the variance of the interest rate var (r(t)). In *case (B)*, (7.5/B), the optimal debt/net worth f* is constant. It is equal to the *mean* return on capital less the *mean* interest rate, (b − r), divided by the variance var [b(t) − r(t)], which contains the covariances. This case is fully analyzed, using dynamic programming, in Fleming and Stein (2004). Case (C), (7.5/C) is the reverse of case (A). The debt/net worth ratio f*(t) varies with time. It is equal to the *mean* return less the *current* interest rate divided by the variance of the return var b(t).

Box 7.2 Optimal Debt/Net Worth Ratio

Case A

$$f^*(t) = [b(t) - r]/\sigma^2{}_r \qquad (7.5/A)$$

$\sigma^2{}_r$ = *variance r(t)*

Case B

$$f^* = (b - r)/\sigma^2 + f(0) \qquad (7.5/B)$$

σ^2 = *variance (b − r)*

Case C

$$f^*(t) = [b - r(t)]/\sigma^2{}_b. \qquad (7.5/C)$$

$\sigma^2{}_b$ = *variance b(t)*

7.6 Basic Data

The optimal debt/net worth f(t) = L(t)/(t) in Box 7.2 depends upon several variables. (1) The return on capital b(t), equal to the sum of the productivity of capital β(t) and the capital gain d(t)/P(t); (2) the interest rate r(t) and (3) risk factors, the variances. Table 7.3 presents the summary statistics of the main variables, and several are graphed in Figs. 7.3 and 7.4. Acronyms are:

Debt/equity ratio DEBTEQUITY = f(t) = L(t)/X(t),
Gross value added/capital = GVACAP = *β(t)* = Y(t)/K(t),
Interest rate = INTDEBT = r(t),
Net return = RETVAINTD1 = GVACAP − INTDEBT = *β(t)* − r(t) = Y(t)/K(t) − r(t),
RET1 = β(t) + dP(t)/P(t) = GVACAP + GROWASSETS.

The value of capital, assets, is K(t) = P(t)Q(t). The growth of assets is d(P(t)Q(t)) where P(t) is the price index of the assets and Q(t) is an index of physical quantity. Since the physical quantity in the farm sector does not vary much one can approximate d(PQ) as K(t)dP(t)/P(t) ~ dP/P, the capital gain corresponds to the growth of assets.

A crucial variable in the optimization process is the difference between the return to capital and the interest rate. The *return to* capital b(t) is the sum of the productivity of capital *β(t)* = GVACAP plus the dP/P the capital gain.

The series in Fig. 7.3 relate *productivity* of capital GVACAP = Y/K = to the interest rate INTDEBT. The series RET1 corresponds to b(t) in (7.2) is a gross return on assets. Figure 7.4 graphs the return to capital RET1 = b(t) and the interest rate INTDEBT = r(t). The correlation ρ between b(t) and the interest rate r(t) is −0.63. When interest rates fall (rise) the capital gain increases (decreases).

7.7 Mean-Variance Interpretation

Figure 7.5 is Mean-Variance interpretation of the expected growth rate for Case B. The derivation and technical details are in Chaps. 4 and 5. A similar diagram can be used for the other cases.

Case B

$$(1/T)\ E\left[ln\ X(T)/X(0)\right] = Mean - Risk = expectedgrowthrate. \tag{7.6}$$

$$Mean = (b - c) + (b - r)f \tag{7.6.1}$$

$$Risk = (1/2)\sigma_b{}^2[f^2\theta^2 + (1+f)^2 - 2f(1+f)\theta\rho] \tag{7.6.2}$$

$$\theta = \sigma_r/\sigma_b,\ \rho = correlation\ between\ (b,r)$$

There are two terms in the expected growth rate equation (7.6). The first term Mean is the straight line in Fig. 7.5/Eq. 7.6.1. It is the expected growth rate for any debt ratio f(t), *if there were no risk.* The intercept is the appropriate return less the ratio C/X of (consumption-dividends-distributed profits)/net worth, when there is no debt. The slope is the appropriate return less the interest rate. The appropriate measure depends upon the relevant case, (A), (B) or (C). The second term in (7.6) is graphed as the parabola Risk in Fig. 7.5/Eq. 7.6.2. It is the variance of the change in net worth.

The expected growth rate of net worth in (7.6) corresponds to the difference between the straight line *Mean* and the parabola *Risk* in Fig. 7.5. In case (B), Mean and Risk are described by (7.6.1) and (7.6.2) respectively.

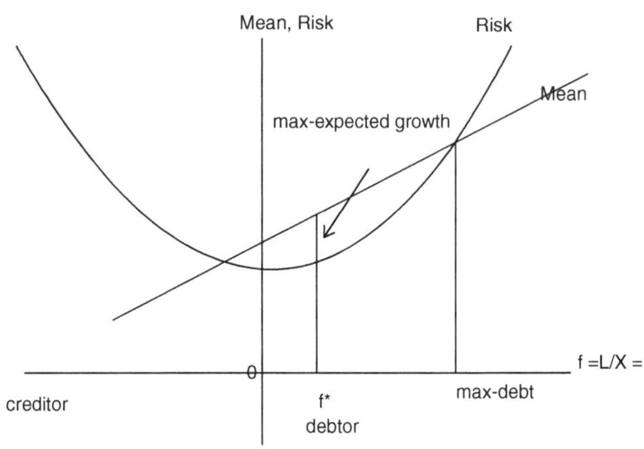

Fig. 7.5 Mean-variance interpretation of expected growth

The optimal debt ratio f* maximizes the distance between Mean and Risk. It is the value in (7.5/A), (7.5/B), or (7.5/C), depending upon the stochastic processes: cases A, B and C. At debt ratio f*, the expected growth rate is maximized for any given ratio C/X of consumption-dividends/net worth. As the debt ratio f = L/X rises above the optimum f*(t), the Risk rises relative to the Mean, and the expected growth rate of net worth declines. At a debt ratio equal to max-debt, the expected growth rate is zero. When the debt ratio rises above max-debt, the expected growth rate is negative and risk is very high. The likelihood of bankruptcy increases continuously as the debt ratio rises above the optimum. The difference $\Psi(t) = [f(t) - f^*(t)]$ is the "excess debt" ratio. From Fig. 7.5, one sees that the likelihood of a serious decline in net worth that threatens bankruptcy is a continuous function of $\Psi(t)$ the excess debt ratio.

7.8 Early Warning Signals

An early warning signal $\Psi(t) = [f(t) - f^*(t)]$ is the "excess debt" ratio. I measure each component, the actual debt ratio f(t) and an optimal ratio f*(t) from Box 7.2. The main component of the optimal ratio is a measure of the "net return", the numerator of the appropriate optimal debt ratio. The appropriate measure of the net return in (7.5/A), (7.5/B), and (7.5/C) depends upon the stochastic process. One cannot be sure what is the appropriate stochastic process and hence optimal debt ratio. In (7.5/A) or (7.5/C), the return should be based upon the current value of one variable and the mean value of the other variable.

My approach is as follows. First, the capital gain term dP/P has a mean that is not significantly different from zero and is stationary. Therefore in (7.5/B), the mean return $b = \beta$, the productivity of capital, is the mean ratio of value added/capital (GVACAP). The numerator of f* should be $(\beta - r)$ = the productivity of capital (GVACAP) less interest rate (INTDEBT), which are graphed in Fig. 7.3 above.

Second, a general approach in evaluating excess debt and obtaining Early Warning Signal is that the optimal debt ratio should follow the net return

$$\text{RETVAINTD} = \text{GVACAP} - \text{INTDEBT} = (b(t) - r(t)). \tag{7.7}$$

In (7.5/A) the appropriate net return is $[\beta(t) - r]$, in (7.5/B) it is $(\beta - r)$ and in (7.5/C) it is $[\beta - r(t)]$. In Fig. 7.6, the curve labeled RETVAINTD is the *normalized* value of $[\beta(t) - r(t)]$ in (7.7).

$$RETVAINTD = [(\beta(t) - r(t)) - (\beta - r)]/\sigma, \tag{7.8}$$

σ = standard deviation of $(b(t) - r(t))$,
$(\beta - r)$ = mean net return

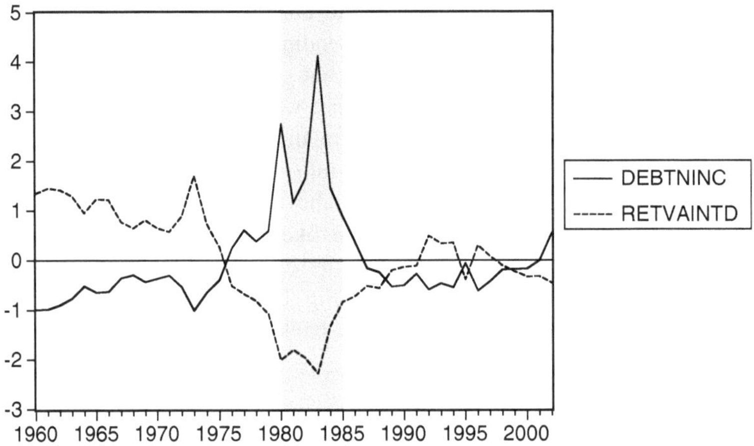

Fig. 7.6 Agriculture. DEBTNINC = L/Y = Debt/net income; RETVAINTD = GVACAP − INTDEBT = (gross value added/assets − interest rate). Normalized variable = (variable − mean)/standard deviation

Third, the debt ratio in the optimization is f = L/X = debt/net worth. However, there is a bias in using this as an empirical measure of an Early Warning Signal (EWS). The reason is that as net worth EQUITY collapses, this ratio jumps up violently. For this reason, in empirical work I prefer to use the ratio h = L/Y of debt (L) to (Y) to net income. Call h the debt ratio. In Fig. 7.6, the normalized value of the *realized* debt ratio is:

$$DEBTNINC = [L(t)/Y(t) - (L/Y)]/\sigma \tag{7.9}$$

σ = standard deviation of [L(t)/Y(t)],
L/Y = mean (L(t)/Y(t))

The optimal debt ratio should either follow RETVAINTD, (Eq. (5A), (5C)) or be constant (7.5/B). My measure of an excess debt $\Psi(t)$ is the difference between the normalized curves in Fig. 7.6. *Non-optimal debt would occur if the debt ratio were rising relative to its long term mean when the net return was declining relative to its mean.*

$$Excess\ debt\ \Psi(t) = DEBTNINC - RETVAINTD > 0 \tag{7.10}$$

In Fig. 7.6, the *normalized* net return fell by almost three standard deviations from 1975 to 1984, but the actual debt ratio rose by about three standard deviations during that period. Figure 7.7 graphs a normalized value of the excess debt $\Psi(t)$. It is the normalized difference between DEBTNINC and (GVACAP − INTDEBT). A large value of normalized deviation $\Psi(t)$ is an EWS of an impending crisis. This crisis did indeed occur, seen in Fig. 7.2, with the bankruptcies and defaults.

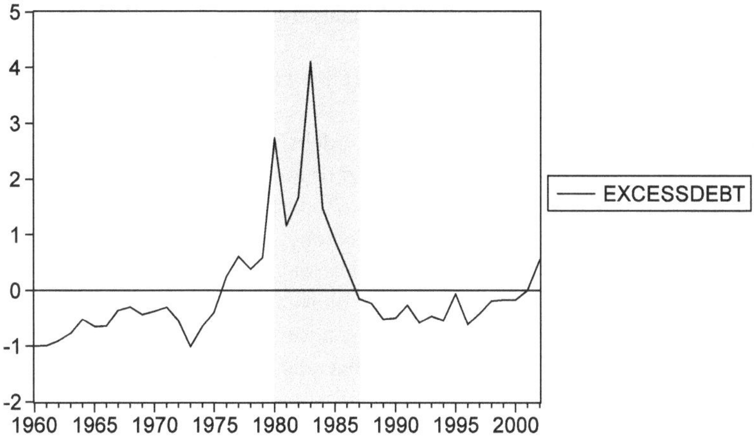

Fig. 7.7 EXCESS DEBT = DEBTNINC − RETVAINTD, normalized: mean zero, standard deviation unity. Shaded period, agricultural crisis

During the periods when $\Psi(t)$ was small, there were no crises. Figure 7.7 summarizes the contribution of the SOC approach to provide an EWS of a debt crisis.

A remarkable result is that the same techniques of analysis, SOC, were used to explain the financial crisis of 2008 and the agricultural crisis of the 1980s. Both were predictable using the techniques of analysis explained in Chap. 4.

7.9 The S&L Crises

The Agricultural debt crisis and the S&L crisis of the 1980s followed similar scenarios. The financial debt crisis of 2008 contained the many of the ingredients of the two financial crises of the 1980s, but involved a national financial structure rather than a regional one. In each case, the growth of the debt was stimulated by initial capital gains on assets that increased equity. The productivity of capital was less than the mean interest rate but the capital gains exceeded the mean interest rate. Borrowing was based upon the rising equity, so debt service/income rose. Capital gains (growth in equity) are not sustainable unless they reflect the growth of the productivity of capital. When the capital gains fell below the interest owed, debts could not be repaid and a crisis occurred.

In 1979, when it was clear that there was a financial crisis, a bailout or rescue plan for the S&L industry was undertaken. Between 1975 and 1995, the rescue plan cost the taxpayers \$153 billion – or \$320 billion in 2011 prices. I draw upon Curry and Shibut (2002) and Sungard Ambit Risk (2002). The question is: how could such a crisis occur in a backwater part of the US financial industry – the S&L industry – that simply takes retail savings deposits and lends them out in the form of residential mortgages? Moreover, this sector was regulated by the Federal Savings & Home Loan Bank Board, which had its own insurance fund to protect depositors, FSLIC.

Equation 7.11 is the framework of the analysis.

$$dX(t) = K(t)[\mathbf{dP(t)/P(t)} + \beta(\mathbf{t})] - \mathbf{r(t)}L(t)\,dt - C(t)\,dt \qquad (7.11)$$

K(t) = P(t)Q(t) = value of assets, L(t) = debt, X(t) = networth = K(t) − L(t), r(t) = interest rate, dP(t)/P(t) = capital gain or loss, C(t) = consumption.

The inflation and oil price shock in 1979–70 pushed up interest rates, which went up as high as 16% on government debt. Thereby, a gap opened up for the S&L's between the cost of funding their liabilities and the income generated by their assets, which were long term, fixed rate mortgages. Term ***[β(t)K(t) − r(t)L(t)]*** declined, where ***β(t)*** is the return on long term fixed rate mortgages, r(t) short term interest rates. Net worth X(t) often became negative – many of the S&L's became insolvent.

New regulations loosened capital requirements and raised the level of insured deposits. With deregulation and removal of interest rate ceilings, S&L's had to compete for funds with the money market funds by offering interest rates r(t) in line with the market rates. Banks could attract funds by raising interest rates, because savers knew that their deposits were insured. Banks used these funds to shift away from conventional mortgages to more risky and higher yielding assets. From the early 1980s S&L's began to lend to real estate developers and to invest in real estate, construction, especially in Texas and Florida. In these regions, property values, real estate prices P(t) were rising due to the oil price boom. In the farm regions, the agricultural farm price boom was another factor inducing bank lending/investment. Large capital gains dP(t)/P(t) were anticipated.

The S&L lending practices had several failings that were repeated by the financial sector in the 2000–2007 period. There was: (a) an overemphasis on the up-front fees from advancing commercial real estate loans; (b) a loosening of underwriting standards which should have ensured creditors that the debtor had sufficient cash flows from the property development; (c) the loan was unduly large relative to the property used as security; (d) untrustworthy property value appraisals that failed to take into account downward movements in property prices in local markets, including purely speculative assumptions about future capital gains.

The debt L(t) in (7.11) rose. During 1970–1977, there were regional crises triggered by the collapses in oil prices, prices and property values in the farm sector. By 1974, the oil price boom was faltering and by 1977 Texas's oil and real estate sectors were in deep recession. A similar process of increasing rates of default and falling collateral values undermined S&L assets around the US until 1992. Texas S&L's were the worst hit. The crisis was due to the negative value of term ***{K(t) [β(t) + dP(t)/P(t)] − r(t)L(t)}***.

Public awareness of the scale of the S&L crisis was limited into the late 1980s, despite the well publicized financial scandals. This changed significantly after President Bush's speech in February 1979. Congress then passed an act that substantially restructured the US financial industry regulation and the US taxpayers would end up paying the bill for the S&L fiasco. The financial market and the Fed in the 2000–2007 period ignored the lessons of the two crises in the 1980s.

References

Barry, Peter J. and Lindon J. Robison. 2001. Agricultural finance: Credit constraints and consequences. In *Handbook of agricultural economics*, eds. B. Gardner and G. Rausser, ch. 10. Amsterdam: Elsevier Science B.V.

Curry, T., and L. Shibut. 2002. The cost of the savings and loan crisis: Truth and consequences. *FDIC Banking Review*.

Economic Research Service USDA. 2002. Agricultural income and finance outlook/AIS-77/ February 26,2002.

Federal Deposit Insurance Corporation (FDIC). 1997. An examination of the banking crises of the 1980s and early 1990s, history of the eighties – lessons for the future, banking and the agricultural problems of the 1980s, ch. 7 <www.fdic.gov/bank>.

Fleming, Wendell H., and Jerome L. Stein. 2004. Stochastic optimal control, international finance and debt. *Journal of Banking & Finance* 28: 979–996.

Moss, Charles and Schmitz, Andrew. 2003. *Governement policy and farmland markets*. Ames: Iowa State Press.

Sungard Ambit Risk. 2002. US Savings & Loan Crisis.

United States Department of Agriculture (USDA). 2002. Economic research service (ERS), farm income and balance sheet indicators, 1929–2002, November 26, 2002.

Chapter 8
The Diversity of Debt Crises in Europe

Abstract The Stability and Growth Pact is seriously incomplete because it only focuses on government deficits and debt but does not focus upon the external net debt arising from the excessive accumulation of private external debt as well as external government debt. The cases of Ireland and Spain differ from Greece, Portugal and Italy. This chapter provides the underlying analytics for understanding the debt crises in Europe by drawing on the NATREX model that explains the simultaneous evolution of the external debt and real exchange rate.

The external debts of the European countries are at the core of the current crises. Generally, the crises are attributed to government budget deficits in excess of the values stated in the Stability and Growth Pact (SGP)/Maastricht treaty. Proposals for reform generally involve increasing the powers of the European Union to monitor fiscal policies of the national governments and increasing bank regulation.

This chapter explains the inter-country differences in the debt crisis in Europe. My basic questions in the European context are: In the various countries, how were "excessive/non sustainable" external debt ratios produced? Were the crises due to government budget deficits/government dissaving or to private investment less private saving? What is the mechanism whereby the actions of the public and private sectors lead to an unsustainable debt burden, defined as an unsustainable ratio of debt service/GDP? The Stability and Growth Pact/Maastricht Treaty and the European Union focused upon rules concerning government debt ratios and deficit ratios. They ignored the problem of "excessive" external debt ratios in the entire economy that led to a crisis in the financial markets. An austerity policy is not necessarily an optimal solution for all of the highly indebted countries. The answers determine to a large extent how one should evaluate proposals for economic reform to avert future crises.

J.L. Stein, *Stochastic Optimal Control and the U.S. Financial Debt Crisis*,
DOI 10.1007/978-1-4614-3079-7_8, © Springer Science+Business Media New York 2012

In the previous chapters the debt ratio was a control variable. Using stochastic optimal control, I derived optimal debt ratios. This is *normative* economics. This chapter is concerned with *positive* economics. The external debt ratio is not a control variable, but is an endogenous variable that is determined by "fundamentals" in a dynamic manner. The "fundamentals" are determined by the actions of both the public and the private sectors. I explain this by drawing upon the Natural Real Exchange Rate NATREX model for the underlying analytics of the equilibrium real exchange rate and external debt – the endogenous variables – that produced the crisis. Although this chapter does not directly concern the U.S. financial crisis nor does it use SOC, the importance and timeliness of the European debt crises induced me to conclude this book with this chapter.

I start by presenting some relevant basic statistics. They strongly suggest the inter-country differences that caused the debt crisis. The government sector was the main cause in Greece and Portugal. The private sector was the main cause in Ireland and Spain.

8.1 Basic Statistics Related to the Origins of the Crises

Table 8.1 presents the Government Structural Balance as a percentage of potential GDP. It refers to the general government cyclically adjusted balance, adjusted for nonstructural elements beyond the economic cycle. Table 8.2 presents the Government Net Debt as a percent of GDP. Both are derived from EconStats, IMF, World Economic Outlook.

Table 8.1 shows the general government structural balances as a percent of potential GDP (SBGDP). The last row contains the mean and standard deviation in the *pre crisis period 1998–2007*. Relative to the Euro area, in Greece and Portugal the SBGDP have been on average twice as high, in Spain the SBGDP have been significantly lower, in Ireland they have been similar and in Italy somewhat higher. A difference between Spain and Ireland is that, from 2001 to 2007, the structural budget deficits in Ireland increased significantly, but were relatively stable in Spain.

Table 8.2 shows that the government debt ratio and hence debt burden – interest payments/GDP – was rising in Greece and Portugal from 1998 to 2007 before the crisis. The ratio of debt burden in the row 2007/1998 was Portugal 1.42, Greece 1.16, Spain 0.53 and Ireland 0.43. One infers from these two tables that the trend in fiscal policy/government budgetary policy – where structural deficits led to debt burdens – was primarily at the origin of the crises in Greece and Portugal, whereas the origin of the crisis in Ireland and Spain (explained below) was primarily the private sector and subsequent government bailout.

When the crisis in the private banking sector occurred in some countries, particularly Ireland and Spain, the government bailed out the banks by purchasing private debt in exchange for public debt. The government/taxpayers then became

Table 8.1 Government structural balance % GDP (SBGDP)

	Euro	Spain	Ireland	Portugal	Greece	Italy
1998	−2.03	−1.736	1.219	−3.4	−2.86	−3.1
1999	−1.6	−1.02	0.269	−3.38	−1.89	−1.8
2000	−1.969	−1.22	1.673	−4.7	−2.68	−0.9
2001	−2.676	−1.757	−1.8	−5.5	−3.647	−3.1
2002	−2.86	−1.1	−2.757	−4.9	−4.1	−3.0
2003	−3.1	−0.976	−3.167	−4.89	−6.03	−3.5
2004	−2.98	−0.978	−2.75	−5.2	−8.638	−3.6
2005	−2.67	−1.598	−3.756	−5.7	−6.01	−4.4
2006	−2.07	−1.275	−4.0	−3.9	−4.9	−3.3
2007	−1.83	−1.132	−7.3	−3.4	−6.795	−1.5
2008	−2.58	−4.9	−11.26	−4.02	−11.47	−2.7
1998–2007	−2.38	−1.28	−2.24	−4.5	−4.76	−2.81
Mean (standard deviation)	(0.54)	(0.31)	(2.71)	(0.09)	(2.11)	(1.02)

Sources: Tables 8.1, 8.2. EconStats, IMF World Economic Outlook; Italy, Federal Reserve St. Louis, International Economic Trends, Government budget balance/GDP

Table 8.2 Government net debt % GDP

	Spain	Ireland	Portugal	Greece
1998	57.37	53.22	42.96	71.73
1999	54.48	47.8	42.01	75.85
2000	50.27	36.67	41.97	77.41
2001	47.54	27.3	46.39	81.15
2002	44	25.13	48.12	84.47
2003	41.3	22.7	51.22	81.93
2004	38.58	19.94	53.28	82.88
2005	34.71	15.92	57.95	84.07
2006	30.53	12.16	58.77	81.66
2007	26.52	12.18	58.1	80.35
2008	30.36	23.04	61.13	83.4
2009	43.73	36.41	72.08	96.83
1998–2007	42.53	27.3	50.08	80.25
Mean (sd)	(10.16)	(14.53)	(6.77)	(4.14)
2009/2007	1.65	2.99	1.24	1.21
2007/1998	0.53	0.43	1.42	1.16

Source: EconStats, IMF World Economic Outlook

the debtor to foreign investors. The effect of the government bailout of the private banking sector is seen in row labeled 2009/2007. It is the government net debt ratio in the post crisis year 2009 relative to its level in the pre crisis year 2007. The difference relates both to the bailout and to the loss of tax revenues from the recession.

8.2 Crises in Ireland, Spain and Greece

I describe the country experiences by drawing upon the reports of the central banks and the EEAG (2011). The story is the same for Ireland and Spain, but different in Greece and Portugal. First, I describe what happened in Ireland and Spain and then in Greece. In Ireland and Spain the structural government budget deficits were lower than in the Euro area (Table 8.1). In these countries, the private sector – housing and banks – was the origin of the debt crisis. Low interest rates, based upon the assumption that there was neither a default risk nor an exchange risk, led to an elastic supply of external funds for the Euro area countries. The demand for non-tradables, in particular housing, was stimulated by low interest rates and growth. Private investment exceeded private saving, the relative price of non-tradables rose, which appreciated the real exchange rate, and thereby generated current account deficits and external debt. In Greece, the current account deficit was attributable to both the government and the private sectors. Later in this chapter, I use the NATREX model to explain the underlying analytics of the trajectories of the external debt and equilibrium real exchange rate, the ratio of domestic/foreign prices. As the external debt burden rises, the probability of a crisis increases.

8.2.1 Ireland

The Celtic Tiger boom in the late 1980s brought sustained growth in employment, income and household formation. Ireland's becoming a founding member of the Euro zone brought a sustained fall in nominal interest rates, which in turn led to higher asset valuations. The growing construction boom was financed by Irish banks which in turn were financed by external financial markets where inexpensive funds were available. In the last 4 years of the boom from 2003 onwards banks competed aggressively in the mortgage markets with little regard for the creditworthiness of the mortgagors. At the end of 2003, net indebtedness of Irish banks to the world was over 10% GDP. By 2008, borrowing mainly for property jumped to over 60% GDP.

Even before the failure of Lehman Brothers (Sep. 2008) Irish residential properties had been falling for more than 18 months. At no point throughout the period, even as the crisis neared, did the Central bank and Financial Services Authority of Ireland staff believe that any of the institutions were facing serious underlying difficulties let alone insolvency problems.

When the crisis occurred, the collapse of construction and the fall in property prices led to the insolvency of banks. Their net worth vanished. The state took large equity stakes in most banks and issued government guaranteed bonds. Although Ireland's public debt immediately prior to the crisis was low (Table 8.2), the fiscal deficit and public sector borrowing surged. The structural budget deficit increased significantly during the period 2000–2007. It went from a surplus of 1.67% in year 2000 to a deficit of 7.3% in 2007. A reason for the surge in the deficit was the collapse of tax revenues in 2008–2009 due to the collapse of the housing sector.

8.2.2 Spain

Throughout the 1989–2006 period, demand in the Spanish economy grew at 4.7% pa whereas output expanded by 3.8% pa, driven by immigration and increased labor force participation. There was scarcely any growth in productivity. Since absorption exceeded production, external debt grew and the real exchange rate – equal to the nominal rate times relative prices – appreciated. Insofar as the nominal exchange rate was fixed in the Euro area, Spanish prices and costs rose relative to the rest of the Euro area. The Spanish economy lost competitiveness. Unlike previous expansions, the resort to financing was not chiefly by the public sector which reduced its debt throughout the period. See Table 8.2. Instead it was the Spanish households and firms that swiftly increased their debts. The real estate market propelled the expansion. Housing prices climbed from an average rate of 1% pa between 1995 and 1997 to 18% pa between 2003 and 2004, or an annual average increase between 1995 and 2007 of 10%.

What facilitated the growth in the debt was the availability of cheap credit in the international financial markets. See EEAG Fig. 4.12 for the spread between Spanish and German 10 year bonds, based upon Reuters Ecowin. As a result, the Spanish economy, which needed virtually no foreign funding in 1996, became a borrower. In 2008 total net borrowing from the rest of the world was 9.1% GDP.

When housing prices fell, the banks – which financed the housing sector – were unable to repay their loans to the international lenders. Governments responded forcefully to the intensification of the financial crisis. At first, measures had focused on the selective bailout of ailing banks, supplementing the actions of central banks to prevent liquidity problems in the banking sector from becoming insolvency problems. However, the risks of financial collapse and the increasingly evident and heightened prospect of global recession led to the widespread approval of plans to support the financial sector and to boost aggregate demand via fiscal stimulus. The breadth of the measures adopted and the volume of resources mobilized were on an unprecedented scale. The government structural deficit in Spain was relatively constant during the period 2000–2007. The effect of the crisis raised the ratio of government debt 2009/2007 to 1.65 (Table 8.2).

8.2.3 Greece's External Debt

The crucial variable leading to a debt crisis is the external debt/GDP. The change in the external debt is the current account deficit. These variables are presented in Tables 8.5 and 8.6 for five countries and the Euro area. Both the government sector and the private sector can produce growth of the external debt. The current account deficit is equal to social investment less social saving (I − S), where social refers to the sum of the private and government sectors. Alternatively, the current account deficit is equal to private investment less private saving ($I - S_p$) plus the government deficit ($G - T$), negative government saving.

$$-CA = (I - S) = \left(I - S_p\right) + (G - T)$$

In the case of Greece, on average from 2000 to 2008, the current account deficit of 10.6% of GDP can be decomposed as

$$[-CA = 10.6\%] = \left[\left(I - S_p\right) = 4.7\%\right] + [(G - T) = 5.9\%].$$

Viewed this way both the private and public sectors accounted for the average current account deficit in Greece. The Greek private sector's saving ratio S_p was an important variable in the decline in the aggregate saving ratio, which contributed to the current account deficits. The private saving ratio declined from 27% in 1988 to 11% in 2008. The decline in the Greek saving ratio is larger than in any other EU country. The growth in government spending in Greece is largely accounted for by the growth in social transfers, which rose from 8% GDP in 1970 to 21% in 2009.

8.3 Role of the Capital Market

The capital market assumed that, since these countries are in the Euro area/common currency, there is neither an exchange rate risk nor a default risk. The capital market treated these countries alike insofar as interest rates were concerned, and did not charge countries a risk premium relative to the rest of the Euro area during the period 2000–2008. Figure 8.1 shows interest rate spreads relative to the German Bund, for Ireland, Greece and Portugal. See EEAG Fig. 4.12 for the spread between Spanish and German 10 year bonds. All are based upon Reuters Ecowin (Plagnol 2010/Reuters).

From the time of the Greek entry in 2000 until 2008 when there was the Lehman Brothers failure, the spread was negligible. The credit default swap (CDS) rate is the premium paid for insurance against default. (A CDS rate of *y* basis points *bp* means that it costs $1,000y to insure $10 million of debt for 5 years.) The CDS rates for Ireland and Greece indicated little doubt about default until 2009. The rates for both were below 100 bp until the beginning of 2009. The CDS rate for Ireland ranged between 100 and 150 bp whereas for Greece the rate rose to 400 bp. Hence no debt problem was anticipated for these countries. Effectively, there was a large supply of international funds at low interest rates to finance the gap between investment and social saving in Euro area countries (Fig. 8.1).

The situation only changed when Lehman Brothers failed. Then the CDS rates and interest rates on Greek and Irish securities rose. The conclusion is that the market for bonds denominated in Euros did not reflect doubts about default until quite late in the crisis. The precipitating factor in the recognition of default risk in

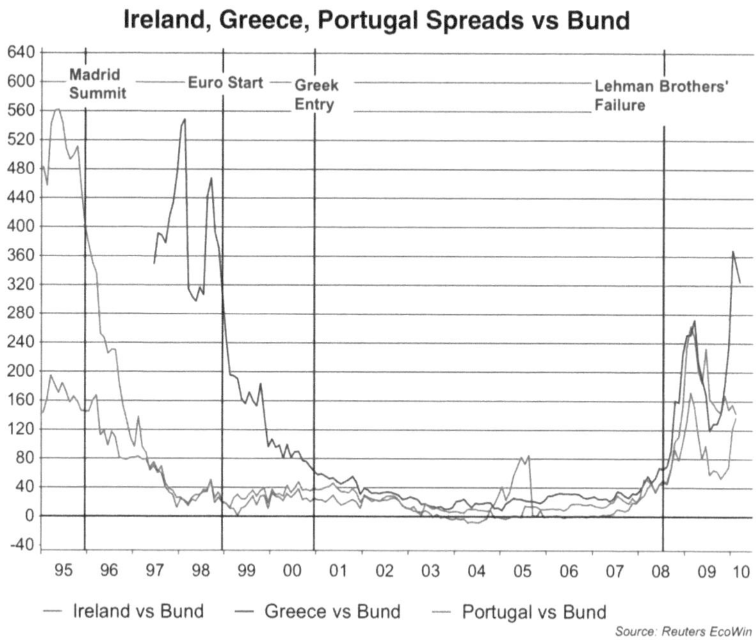

Fig. 8.1 Interest rate spreads versus the Bund

Table 8.3 United State's municipal rating distribution 1970–2000

Bond rating	Percent (%) of municipalities
AAA	3.15
Aa	11.5
A	54.42
Baa	29.9
Ba	0.8
B	0.13
Caa-C	0.02

Special comment: Moody's U.S. Municipal Bond Rating Scale, Nov. 2002

Europe was the failure of Lehman Brothers. By then it was too late. C. Fischer found that the rescue of Bear Stearns in March 2008 seems to mark a change in market perceptions of sovereign bond risk. The government bonds of some countries lost their former role as a safe haven.

The ignoring of default risk stands in contrast with the U.S. experience where, despite having a common currency, the market evaluates municipalities according to the default risk. Neither the Treasury nor the Federal Reserve intervenes in the fiscal policies of the municipalities nor contemplates bailouts when they are experiencing difficulties in servicing their debts. Table 8.3 presents the distribution of ratings, and hence interest rates that the capital market charges the various U.S. municipalities. *In the U.S. the fiscal discipline comes from the markets and not from the government.*

Table 8.4 Banks and governments: debtor, creditor by country, $billions

Debtors	Greece	Ireland	Italy	Spain	Portugal	Britain	France	Germany	Total debt
Greece	–	8.5	6.9	1.3	9.7	15	75	45	236
Ireland	0.8	–	18	16	22	188	60	184	867
Italy	0.7	46	–	47	5.2	77	511	190	1,400
Spain	0.4	30	31	–	28	114	220	238	1,100
Portugal	0.1	5.4	6.7	86		24	45	47	286
Britain									
France									
Germany									
Total	2	89.9	62.6	150.3	64.9	418	911	704	

Source: Fidelity Investments, Strategic Advisers, 2010
Row is debtor and column is creditor

8.4 Repercussions in Financial Markets

It is difficult to separate bank debt from government debt when the governments have bailed out banks. The government/taxpayer takes over the role of the debtor. There is reason to combine the two debtors. Table 8.4 displays the debts of the banks and governments.

Debtor is listed in row and creditor in column. The major debtors were Italy, Spain and Ireland. Thus Spain owed $220 billion to the French and $238 billion to the Germans. The major creditors were the French, German and British banks. The major creditors for Ireland were Britain and Germany. Greece and Portugal were minor debtors among the European countries. Last column is total debt to all countries in addition to those in the table.

When the crises occurred in Greece, Portugal, Ireland and Spain, whether due to the government or the private sector, defaults occurred or were threatened. If Spain defaulted then assets of the British, French and German banks/government declined in value. If the Irish defaulted, the British and German banks/government were affected. If Italy defaulted, the French and German banks would be affected. This set of linkages was a source of concern to the ECB.

8.5 Origins of the External Debt Ratio

The relevant debt is the external debt, and a crisis occurs when the debt service payments/GDP are unsustainable. The net external debt is the sum of current account deficits. The internal government debt/GDP per se does not affect the vulnerability of an economy and does not produce debt crises.

Table 8.5 Current account/GDP

Obs	Euro area	Greece	Ireland	Spain	Italy	Portugal
1998		−2.8	0.8	−1.2	1.6	−6.8
1999	−0.59	−5.6	0.3	−2.9	0.7	−8.8
2000	−1.35	−7.8	−0.4	−4.0	−0.5	−10.4
2001	−1.06	−7.3	−0.6	−3.9	−0.1	−10.3
2002	0.66	−6.8	−1.0	−3.3	−0.8	−8.23
2003	−0.08	−6.5	0.0	−3.5	−1.3	−6.4
2004	0.78	−5.8	−0.6	−5.3	−0.9	−8.3
2005	0.02	−7.6	−3.5	−7.4	−1.7	−10.3
2006	−0.56	−11.3	−3.6	−9.0	−2.6	−10.7
2007	0.13	−14.5	−5.3	−10.0	−2.4	−10.4
2008	−1.83	−14.7	−5.6	−9.7	−2.9	−12.6
2009	−1.24	−11.4	−3.0	−5.5	−2.1	−10.2
2010	−0.81	−10.5	−0.3	−5.5	−3.5	

Source: Federal Reserve St. Louis, International Economic Trends; Portugal, World Bank

Table 8.6 External debt position end 2009

Country	Net external debt/GDP	General government net external debt/GDP
Portugal	88.6%	74.4%
Greece	82.5	78.9
Spain	80.6	47.3
Ireland	75.1	70.6
Italy	37.3	42.9
Germany	−21.7	48.5

Source: Ricardo Cabral, The PIGS' external debt problem, VOX, May 8, 2010
Negative value is creditor

8.5.1 Current Account/GDP and Net External Debt/GDP

Table 8.5 describes the current account/GDP in the Euro area and the five countries. The net external debt is the sum of current account deficits. The steady trend in current account deficits led to a growing external debt.

Table 8.6 describes the net external debt/GDP and the general government external debt/GDP, as of the end of 2009. The *net* external debt/GDP in column 1 is the key to understanding the crises in Europe. It is equal to the debt like securities liabilities less the corresponding asset claims. Column 2 is the general government net external debt/GDP. The levels and trends in the current account deficits were greatest in Greece and Portugal.

Looking at Germany one sees the difference between the two measures. Germany is a net *creditor* concerning the net external debt, whereas it has a general government net debt/GDP similar to Spain and higher than Italy. Portugal, Greece and Spain have the highest net external debt/GDP, which combines the private and public sectors. Italy does not have a high net external debt/GDP. Therefore, the

Table 8.7 Residential property prices EU countries, annual % change, new and existing houses

	Germany	Ireland	Greece	Portugal	Spain	Italy	France	Euro area
1996	−1.1	–	9.9	1.7	1.4	2.4	–	2.0
1997	−1.9	–	8.2	3.6	2.8	3.4	0.1	2.3
1998	−1.6	22.6	14.4	4.5	5.8	−1.4	1.9	2.5
1999	1.4	22.5	8.9	9	7.7	0.8	7.1	4.9
2000	0.2	20.5	10.6	7.7	8.6	3.9	8.8	6
2001	0.2	14.0	14.4	5.4	9.9	6.0	1.9	5.5
2002	−1.9	6.1	13.9	0.6	15.7	12.6	8.3	6.8
2003	−1.2	14.3	5.4	1.1	17.6	7.2	11.7	6.4
2004	−1.4	11.5	2.3	0.6	17.4	7.0	15.2	7.2
2005	−1.5	7.2	10.9	2.3	13.4	8.6	15.3	7.6
2006	0.3	13.4	12.2	2.1	10.4	5.8	12.1	6.4
2007	0.3	0.9	–	1.3	5.8	5.0	6.1	4.3
Mean	−0.68	13.3	10.1	3.3	9.71	5.11	8.05	5.16
St.dev.	1.1	7.23	3.8	2.8	5.43	3.7	5.27	1.97

Source: BIS Housing Statistics IFC Bulletin #31 Annex 1

exclusive focus upon the government sector is misleading. Country experiences were different.

The crucial variable leading to a debt crisis is the net external debt/GDP. It is the sum of current account deficits. The accounting identity is: current account = (private saving less private investment) + government saving. The first term in parenthesis refers to the private sector and the second to the government sector. The private sector played an important role in generating an external debt. Low world rates of interest and high domestic growth led to a rise in housing prices. In the period 1991–2000 the growth rates in Ireland and Spain were very high. Table 8.7 indicates the large capital gains resulting from investment in housing/non-tradables in Ireland and Spain, relative to the Euro area. The mean capital gain was Ireland 13.3%, Spain 9.71% and the Euro area 5.16%. The anticipated return was the marginal product of capital plus the anticipated capital gain. Interest rates were low and similar among the countries in the Euro area. The large gap between the return to investment in housing/nontradables and the interest rate generated a demand for nontradables. Irish and Spanish banks borrowed abroad at low rates of interest and loaned these funds to the housing industry. The demand for non-tradables/housing rose, appreciating the real exchange rate – the ratio of domestic/foreign prices – inducing a current account deficit financed by capital inflows and the external debt burden rose. The *exclusive* focus of the SGP upon the government debt and deficit is misleading – as the recent crises indicated. The case of each country is different. Section 8.7/Table 8.9 evaluate the relative importance of the government budget and the private sector demand for non-tradables in Greece, Ireland, Spain, Portugal and Italy in accounting for the external debt.

Table 8.8 NATREX dynamics of exchange rate and external debt: two basic scenarios

Scenarios	Medium run	Longer-run
R = real exchange rate = domestic/foreign prices, rise is appreciation, F = external debt/GDP; initial period T = 0, medium run T = 1, long-run T = 2	T = 1	T = 2
Populist	Appreciation	Depreciation
Rise in social in social consumption (time preference), rise in high employment/structural government budget deficit, decline social saving, rise in demand for non-tradables	R(1) > R(0) Debt rises F(1) > F(0)	R(2) < R(0) < R(1) Debt rises F(2) > F(1) > F(0)
Growth oriented	Appreciation	Appreciation
Rise in productivity of investment, expansion of production of tradables. Rise in growth, rise in competitiveness – increase in trade balance function	R(1) > R(0) Debt rises F(1) > F(0)	R(2) > R(1) > R(0) Debt declines F(2) < F(0) < F(1)

The external debt is an endogenous variable that depends upon "economic fundamentals". In the next part I present the Natural Real Exchange Rate Model (NATREX) concerning the simultaneous evolution of the endogenous variables: the real exchange rate – the ratio of domestic/foreign prices – and the external debt. The "fundamentals" are: (a) the expenditures on non-tradables, that may arise from either the structural government budget balance or expenditures by the private sector such as occurred in the house price bubble; (b) the productivity of the economy that increases the current account, by increasing the output of tradables. In the case (a) the external debt rises from the medium to the long run; and the real exchange rate first appreciates and then depreciates below its initial level. In case (b), the real exchange rate appreciates from the medium to the longer run. The external debt first rises and then declines steadily in the long run to a level below the initial value.

I apply the NATREX model to explain the diversity of debt crises in the European countries. In the empirical part, I explain the relative roles of the different fundamentals in the various European countries.

8.6 NATREX Model of External Debt and Real Exchange Rate

The NATREX model provides the analytic structure for understanding the debt crisis.[1] It concerns the *equilibrium exchange rates and external debt*. The equilibrium real exchange rate and external debt interact in a dynamic manner. The real

[1] The NATREX model is developed in Stein (2006), which contains the derivations and applications.

rate is defined as the nominal exchange rate times relative domestic/foreign prices. In a monetary union, the nominal exchange rate is fixed at unity and the actual real exchange rate varies due to changes in relative prices. In a floating exchange rate regime, both the nominal exchange rate and relative prices can lead to adjustments in the actual real rate. The only difference from our point of view is that the adjustment of the actual real exchange rate to the equilibrium will be faster, and with less adverse effects upon growth, when the exchange rate floats, because the nominal exchange rate is more flexible than relative prices.

The concept and measure of the equilibrium exchange rate depends upon the time horizon and the underlying model. Several reasons have been cited in the literature why it is important to estimate equilibrium exchange rates. *First*, there are significant and sustained movements in exchange rates. These movements affect the competitiveness of the economies and their macroeconomic stability. One wants to know whether these movements are ephemeral or whether they are responding to "real fundamentals". This information is important because the answer has implications for rational macroeconomic policy. If the depreciation of an exchange rate is due to a depreciation of its equilibrium value, then exchange market intervention or a restrictive monetary policy designed to offset the depreciation is counterproductive.

Second, in the case of monetary unions such as the Euro area, it is important to know how a potential entrant should select its exchange rate. An "overvalued" rate will depress growth, increase the external debt and produce problems such as beset the eastern part of Germany. An "undervalued" rate will generate inflationary pressures. The measure of "over-valuation" and "under-valuation" must contain an explicit measure of an "equilibrium" real exchange rate.

The literature of "equilibrium exchange rates" was given great impetus by John Williamson's influential book (1994). The logic of this approach goes back to Ragnar Nurske's article. The "equilibrium" exchange rate is the exchange rate that is associated with both external and internal balance. Anticipations, speculative capital movements and changes in reserves are excluded from the concept of an *equilibrium* exchange rate, which is where the exchange rate is heading. The NATREX model of equilibrium exchange rates generalizes the work of Williamson and Nurske. It is a Neoclassical growth model, whose underlying equations are based upon an intertemporal optimization by the private sector, but not the government whose decisions are political.

The NATREX explains the fundamental determinants of the medium run equilibrium and the dynamic trajectory of the real exchange rate and the external debt to the long run equilibrium. In the medium run equilibrium there are both internal and external balances. In both the medium run and longer run the NATREX equilibrium real exchange rate satisfies Eq. 8.1, subject to constraints. The *constraints* are that there is *internal balance*, where the rate of capacity utilization is at its longer term mean, and *external balance* where the real rates of interest at home and abroad are equal, there are neither changes in reserves, nor speculative capital flows based upon anticipations. The equilibrium real exchange rate is the mean of a distribution, which is based upon real fundamentals. The mean will vary over time due to endogenous changes in capital and external debt, as well as changes in the exogenous real fundamentals. Deviations from this mean are produced by speculative factors involving anticipations, cyclical factors, lags in adjustment, and interest rate differentials. These disequilibrium elements average out to zero. These deviations produce considerable variation but their effects are ephemeral.

All of the authors who take the equilibrium real exchange rate approach use Eq. 8.1 to determine the exchange rate. Investment less saving ($I_t - S_t$) plus the current account is equal to zero. Investment less saving is the non-speculative capital inflow. The current account ($B_t - r_tF_t$) is the trade balance B_t less transfers of interest and dividends r_tF_t. The net external debt is F_t and r_t is the "interest/dividend" rate. The international investment position consists of equity, portfolio investment and direct investment. The debt F_t is the negative of the net international investment position. Measure investment, saving and the debt as fractions of the GDP.

$$[(I_t - S_t) + (B_t - r_tF_t)] = 0 \tag{8.1}$$

The main differences among authors concern their treatment of the two terms. Some work with a concept of what is a "sustainable" current account such that the debt does not "explode". Their estimates are subjective, so their equilibrium exchange rate is an arbitrary normative measure. The NATREX approach is quite different in several respects. First, it is positive economics. The "fundamentals" determine the trajectories of the external debt and equilibrium real exchange rate. Second, it is a dynamic analysis. The endogenous current account generates an evolving external debt, which feeds back into the medium run equation 8.1. A trajectory to longer run equilibrium is generated.

The dynamics of the debt/GDP ratio F_t is Eq. 8.2, where g is the growth rate. The current account deficit is the change in the external debt. The real exchange rate affects the trade balance B in Eq. 8.1, and the trade balance affects the evolution of the actual debt ratio in Eq. 8.2. There is a dynamic interaction between the endogenous real exchange rate and debt ratio.

$$dF_t/dt = (I_t - S_t) - g_tF_t = (r_tF_t - B_t) - g_tF_t = (r_t - g_t)F_t - B_t \tag{8.2}$$

In *longer run equilibrium*, the debt ratio stabilizes at a value that satisfies Eq. 8.3. The trade balance B_t is sufficient to finance the interest plus dividend transfer on the debt net of growth ($r_t - g_t$)F_t. A negative debt is net foreign assets.

$$(r_t - g_t)F_t - B_t = 0. \tag{8.3}$$

The *longer-run equilibrium* real exchange rate R_t* and debt/GDP ratio F_t* are endogenous variables that satisfy both Eqs. 8.1 and 8.3. They are written as (8.4) and (8.5) to indicate that they both depend upon the real fundamentals Z_t.

$$R_t* = R(Z_t) \tag{8.4}$$

$$F_t* = F(Z_t). \tag{8.5}$$

The dynamic stock-flow model equations 8.1, 8.2 and 8.3 is the NATREX model, which is an acronym for the *Nat*ural *R*eal *Ex*change Rate. This is a model of *positive economics*. The derivation of the underlying equations is in Stein (2006, Chap. 4). The literature associated with Williamson's FEER uses Eq. 8.1, is normative and does not contain the dynamic interactions, Eqs. 8.2 and 8.3.

8.6.1 Populist and Growth Scenarios

The NATREX model is a technique of analysis. The purpose of the model is to understand the effects of policies and external disturbances upon the trajectories of the equilibrium real exchange rate R_t and equilibrium external debt ratio F_t, which depend upon the vector of fundamentals Z_t. Insofar as the fundamentals vary over time, the equilibrium real exchange rate and debt ratio will vary over time, as indicated in Eqs. 8.4 and 8.5. The logic and insights of the NATREX model can be summarized in two scenarios. Each scenario concerns different elements in the vector Z_t of the fundamentals, and has different effects upon the *equilibrium* trajectories of the real exchange rate NATREX and of the external debt. NATREX analysis concerns the *equilibrium* real exchange rate and it is neither the *actual* real exchange rate nor the *optimal* exchange rate that would lead to the *optimal* debt ratio. Table 8.8 summarizes the differences between the two scenarios in the medium and the long run.

The first scenario, called the *Populist scenario*, involves a decline in the ratio of social saving/GDP. This could occur when the government incurs high-employment budget deficits, lowers tax rates that raise consumption, or offers low interest rate loans for the production of non-tradable goods. This represents a rise in the consumption ratio/decline in the saving ratio, a shift in the S function in Eqs. 8.1 and 8.2. These Populist expenditures are designed to raise the standards of consumption/quality of life for the present generation.

The second scenario, called the *Growth scenario*, involves policies designed to raise the productivity of capital, increase the competitiveness of the economy, and increase the supply of traded goods. Growth policies improve the allocation of resources and bring the economy closer to the boundary of an expanding production possibility curve.

The stories behind the dynamics are as described by Figs. 8.2, 8.3 and Table 8.8. Curve SI in Fig. 8.2 is saving less investment (S − I)(t). It is positively related to the real exchange rate, because a rise in domestic/foreign prices adversely affects investment (see Stein 2006, Chap. 4). The curve labeled CA(t) is the current account function $(B_t - r_tF_t)$, which is negatively related to the real exchange rate, because a rise in domestic/foreign prices adversely affects the trade balance. Initial equilibrium is R = R(0) and CA = 0.

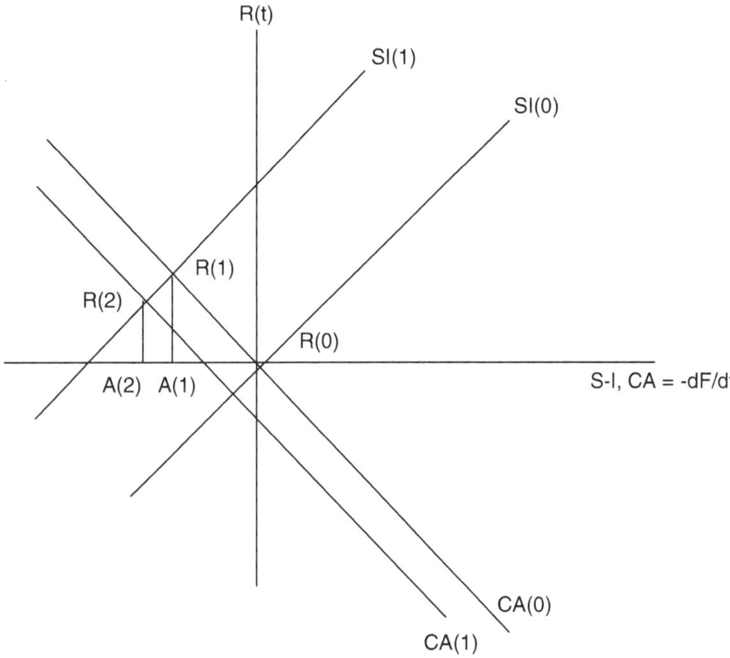

Fig. 8.2 Saving less investment is SI and current account is CA. Decline in social saving shifts SI from SI(0) to SI(1). Real exchange rate appreciates to R(1) and current account declines to A(1). The resulting rise in debt shifts CA from CA(0) to CA(1). Real exchange rate depreciates to R(2) and current account deficit rises to A(2)

The Populist scenario involves decreases in social (public plus private) saving/ increases in social consumption, relative to the GDP. Say that it involves an increase in demand for non-tradables such as housing. External borrowing must finance the difference between investment and saving. The SI function shifts from SI(0) to SI(1). The new equilibrium is at [R(1), A(1)], where T = 1 denotes *medium run equilibrium*. The real exchange rate appreciates because the price of non-tradable goods rises. The price of tradable goods is determined in the world market. The current account deficit is balanced by the capital inflow, now equal to A(1). The debt rises, since the current account deficit is the rate of change of the debt – Eq. 8.2. Current account deficits lead to growing transfer payments r_tF_t.

The rise in the debt decreases the current account function, shifts the curve CA from CA(0) to CA(1). This Populist scenario is potentially dynamically unstable because the increased debt raises the current account deficit, which then increases the debt further. The exchange rate then depreciates to R(2), and since the current account deficit has risen to A(2). The debt rises *steadily*. The populist scenario – a rise in the demand for non-tradables – is described in Fig. 8.3 and Table 8.8. The real exchange rate first appreciates and then depreciates below its initial level. The external debt rises steadily.

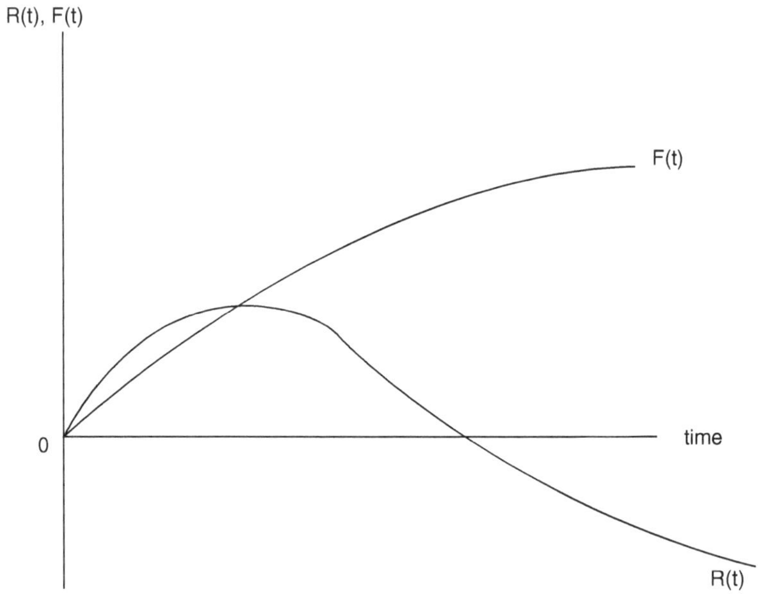

Fig. 8.3 Populist scenario: initial R(0), F(0) at origin. Rise in social consumption, increase demand for non-tradables generates trajectory R(t) for the real exchange rate and F(t) for the external debt. In the Growth scenario, the trajectories for the real exchange rate and external debt trajectory are reversed

Stability can only occur if the rise in the debt, which lowers net worth equal to capital less debt, reduces social consumption/raises social saving. Thereby, saving less investment rises. *Long-run equilibrium* (denoted by $T = 2$) is reached at a higher debt $F(2) > F(0)$ and a depreciated real exchange rate $R(2) < R(0)$. The longer-run depreciation of the exchange rate $R(2) < R(0)$ can be understood from Eq. 8.3. The debt is higher than initially. Therefore, the trade balance B(2) must be higher than initially to generate the foreign exchange to service the higher transfers $r_tF(2)$. The mean interest rate must exceed the mean growth rate if the expected present value of future income is finite. The real exchange rate must depreciate to $R(2) < R(0)$ in order to raise the trade balance to B(2).

The *Growth scenario* is summarized in the lower half of Table 8.8. The perturbation is a rise in the productivity of investment and an expansion of the production of tradables. Investment rises because of the rise in the rate of return. The difference between investment and saving is financed by a capital inflow. S-I shifts from S(0) to S(1) in Fig. 8.2. The exchange rate appreciates to $R(1) > R(0)$ which reduces the trade balance and initially produces a current account deficit A(1). The current account deficit raises the debt. The trade deficit provides the resources to finance capital formation, which raises the growth rate and the competitiveness of the economy.

The investment in the tradables sector increases the *trade balance function* $B = B(R;Z)$. The B function, which relates the value of the trade balance to the real exchange rate R, increases with a rise in the overall productivity/competitiveness of the economy. For example, the reallocation of resources leads to the production of higher quality/value goods that can compete in the world market.

The medium run effects are the same as those in the populist scenario, but the trajectories to longer-run equilibrium differ. The crucial aspect implied by the Growth Scenario is that the economy is more competitive. At exchange rate R (1), *the trade balance function CA increases, shifts to the right.* The real exchange rate appreciates and there are now current account *surpluses*, excess of saving over investment. As a result, the debt then declines to a new equilibrium F(2) < F(0). The trajectory of the debt is not monotonic. The net effect in the longer-run can be understood from Eq. 8.3. The debt is lower, the growth rate is higher and the trade balance function B has shifted to the right. *The long-run equilibrium exchange rate must appreciate to reduce B to equal the lower value of* $(r - g)F^*$.

The dynamic process in the Growth scenario is summarized in the lower half of Table 8.8. The *real exchange rate* appreciates steadily to a higher level R(2) > R(1) > R(0). The *external debt* reaches a maximum and then declines to F(2) < F(0) < F(1). In Fig. 8.3, the curves are reversed.

8.7 NATREX Analysis of the European Situation

I analyze the European experiences within the framework of the NATREX model. Summary Table 8.9 shows that over the period 1998–2010, there were very large current account deficits in Greece, Portugal and Spain, relative to the Euro area. Thereby the external debts of these countries rose. What produced the current account deficits? They were due to what the NATREX model calls the Populist Scenario: a decline in government saving and/or a rise in demand for non-tradables.

The role of the government sector is described in the row labeled government balance. The largest government deficits were in Greece, Portugal and Italy. Hence one concludes that in Greece, Portugal and Italy, the current account deficits were significantly produced by high government consumption/low government saving.

In Spain and Ireland, the government sector was not the major cause of a decline in social saving/rise in social consumption. It was the increase in the demand for non-tradables/housing. This is reflected in the row labeled House Price appreciation. In Ireland, Spain and Greece, there was a significant rise in the demand for non-tradables relative to the Euro area.

The movement of the real exchange rate, equal to the ratio of domestic prices relative to the Euro area, is reflected in the row labeled GDP deflator. In all of the countries, both the government sector and the rise in the demand for non-tradables by the private sector led to an appreciation of the real exchange rate, current account deficits and the growth in the external debt ratio dF_t/dt. The SI curve shifted from SI(0) to SI(1) in Fig. 8.2 changing the medium run equilibrium to

Table 8.9 Summary data 1998–2010

	Euro area	Greece	Ireland	Spain	Portugal	Italy
Current account/GDP	Mean m = −0.26	m = −8.6	m = −1.75	m = −5.4	m = −9.45	m = −1.7
	St. dev. = 0.85	sd = 3.5	s = 2.16	s = 2.8	s = 1.78	s = 1.47
Gov't balance	m = −2.38	m = −4.76	m = −2.24	m = −1.28	m = −4.5	m = −3.16
	s = 0.54	s = 2.11	s = 2.7	s = 0.31	s = 0.09	s = 1.27
House price appreciation, percent change	m = 5.16	m = 10.1	m = 13.3	m = 9.71	m = 3.3	m = 5.11
	s = 1.97	s = 3.8	s = 7.24	s = 5.43	s = 2.8	s = 3.7
GDP deflator, percent change	m = 1.7	m = 3.23	m = 2.45	m = 3.1	m = 2.67	m = 2.35
	s = 0.59	s = 0.84	s = 3.21	s = 1.35	s = 1.04	s = 0.72
Growth, percent change	m = 1.6	m = 2.7	m = 4.2	m = 2.7	m = 1.52	m = 0.72
	s = 2.0	s = 2.8	s = 5.1	s = 2.4	s = 2.06	s = 2.14

Source: Federal Reserve St. Louis, International Economic Trends, Tables 8.1, 8.3

Table 8.10 GDP deflator, percent change from year ago

	Euro area	Greece	Ireland	Portugal	Spain	Italy	U.S.
1998	1.0	5.2	6.6	3.79	2.5	2.7	1.1
1999	1.6	3	4.1	3.29	2.6	1.8	1.5
2000	1.3	3.4	6.1	3.24	3.5	1.9	2.2
2001	2.4	3.1	5.5	3.57	4.2	2.9	2.3
2002	2.6	3.4	4.5	3.73	4.3	3.3	1.6
2003	2.2	3.9	2.8	3.0	4.1	3.1	2.2
2004	1.9	3	2	2.46	4	2.6	2.8
2005	2.0	2.8	2.5	2.51	4.3	2.1	3.3
2006	1.9	3.1	3.8	2.77	4.1	1.8	3.3
2007	2.4	3	1.1	3.18	3.3	2.6	2.9
2008	2.1	3.5	−1.4	1.58	2.4	2.8	2.2
2009	1.0	1.3	−4	0.545	0.6	2.3	0.9
2010	0.8	3.3	−1.7	0.98	0.4	0.6	1

Source: Federal Reserve Bank St. Louis, International Economic Trends, Eurostat

[R(1), A(1)] from [R(0), A = 0]. The resulting change in the external debt is expressed in:

$$dF_t/dt = (I_t - S_t) - g_t F_t = (r_t F_t - B_t) - g_t F_t = (r_t - g_t) F_t - B_t \tag{8.2}$$

The appreciation of the real exchange rate from R(0) to R(1), measured by the GDP deflator, was higher in all five countries relative to the Euro area.

The debt ratio stabilizes if the trade balance B(R(t);Z(t)) is sufficiently large to earn the income to service the debt, adjusted for growth. Since the appreciation of the real exchange rate, the percent change in the GDP deflator (relative to the Euro), was higher in all five countries, the trade balance could not rise to service the debt. A necessary condition for the debt ratio to stabilize is that the real exchange rate depreciates to increase the trade balance (8.3).

Table 8.10 shows that, in the last few years, there has been some real exchange rate appreciation in Greece, but it has remained relatively stable in the other countries. From Table 8.9, one sees that – over the entire period 1998–2010 – Italy has had current account deficits similar to Ireland, less than in Greece, Spain and Portugal and higher than in the EU area. The real exchange rate of Italy appreciated as a result which led to a rise in the external debt. In recent years, the Italian current account deteriorated, Table 8.5.

8.8 Conclusions

I use the NATREX model to explain the causes of the rise in the external debt generated by the current account deficit. They were produced by both *structural* government budget deficits and the rise in the demand for non-tradables, which in

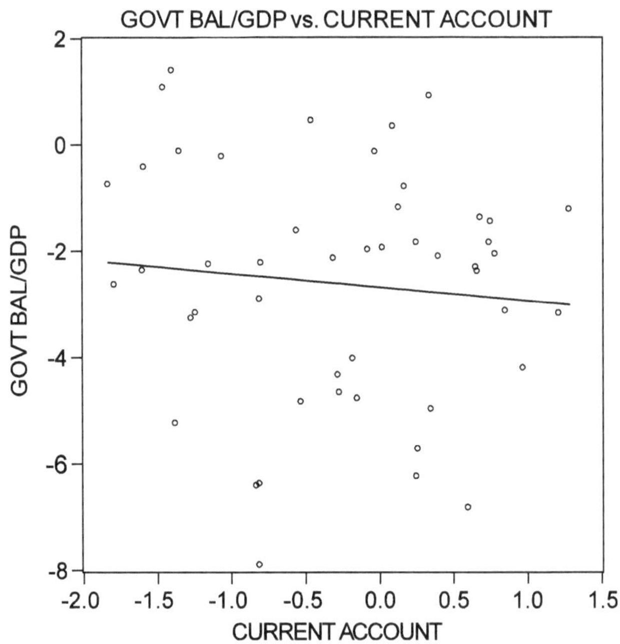

Fig. 8.4 Euro area. Scatter diagram and regression line. Government budget balance/GDP and current account/GDP (*Source*: ECB Statistical Data Warehouse)

turn appreciated the real exchange rate. This is the movement to the medium run equilibrium R(1), A(1) in Fig. 8.2.

The SGP just focused upon the actual/realized budget deficit/GDP and assumed that it was the factor that produced current account deficits. Hence, to avoid an unsustainable debt, policy should be based upon realized budget deficits. This approach is misleading for several reasons. First, the crucial variable in generating crises is the external debt, not the internal debt. Second, one relevant variable for the external debt in the NATREX model is the structural government budget balance, the data in Table 8.1.

Third, *realized budget deficits per se do not have a 1–1 correspondence with current account deficits*. The IMF analysis (IEO 2011, Fig. 8.4) showed that the U.S. current account balance closely tracked the saving-investment balance of households, while the fiscal balance showed little correlation. A similar situation is true for the Euro area. Figure 8.4 is a scatter diagram relating the current account/ GDP to the realized government balance/GDP, for the entire Euro area. Surpluses are positive and deficits are negative. There is not a positive relation between the two.

There are several reasons for the absence of a 1–1 relation between realized budget deficits and current account deficits. (a) The realized budget deficit, call it

GT* = structural budget deficit + cyclical component. Hence GT* has a cyclical component that is affected by the deviation of actual output from capacity output. (b) The current account is also affected by the private sector difference between investment and saving ($I - S_p$). (c) A high ($I - S_p$) raises income, raises the current account deficit, raises tax receipts and lowers the realized government budget deficit. In this case, the negative relation ($I - S_p$) and GT* leads to the absence of a positive relation between GT* and the current account deficit.

The SGP rules must be viewed in the context of the NATREX model above. The policy objective is generally to achieve a reduction in the external debt. No *sustained* improvement in the external debt can occur unless the growth scenario occurs. Growth and the trade balance function must increase. In the shorter run, the real exchange rates of the five countries must depreciate relative to the Euro. Bailouts and austerity policies will be ineffective in reducing the growth of the debt unless the trade balance function B(R(t)) increases relative to the debt service (r − g)F(t). The right hand side of Eq. 8.2 must decline.

$$dF_t/dt = (I_t - S_t) - g_tF_t = (r_tF_t - B_t) - g_tF_t = (r_t - g_t)F_t - B_t \tag{8.2}$$

In the NATREX model, output is at capacity Y(t) = Y*(t), where Y(t) is actual output and Y*(t) is capacity output. The trade balance function in the NATREX model is B(R(t), Y*(t);Z(t)). The equality CA = S − I is brought about by variations in the real exchange rate, as shown Fig. 8.2. If the real exchange rate R(t) = R, ratio of domestic/foreign prices, is constant at R, then the variable of adjustment must be Y(t). An austerity policy per se that raises social saving (decrease government expenditures, raise taxes) will reduce GDP and induce a decline in imports. Thereby B(t) rises because the decline in Y(t) below capacity output reduces imports. But the austerity policy will neither induce growth of real GDP, nor a sustainable rise in the trade balance function when GDP recovers. It therefore will not be effective in reducing the growth of the debt in Eq. 8.2. Austerity policies to be effective must produce the growth scenario.

The *equilibrium* value of the Euro per se has not necessarily been adversely affected by the debt crises of the five countries. The *actual* value of the $US/Euro exchange rate is plotted in Fig. 8.5. The equilibrium value NATREX of the Euro-$US depends upon the fundamentals, the two scenarios, in both the U.S. and Euro area, not just in just one. To be sure there are large variations around the equilibrium value due to speculative and cyclical factors. They average out to a mean of zero. See Stein (2006, Chap. 5) for both a theoretical analysis and estimate of the NATREX for Euro-$US in a two country setting.

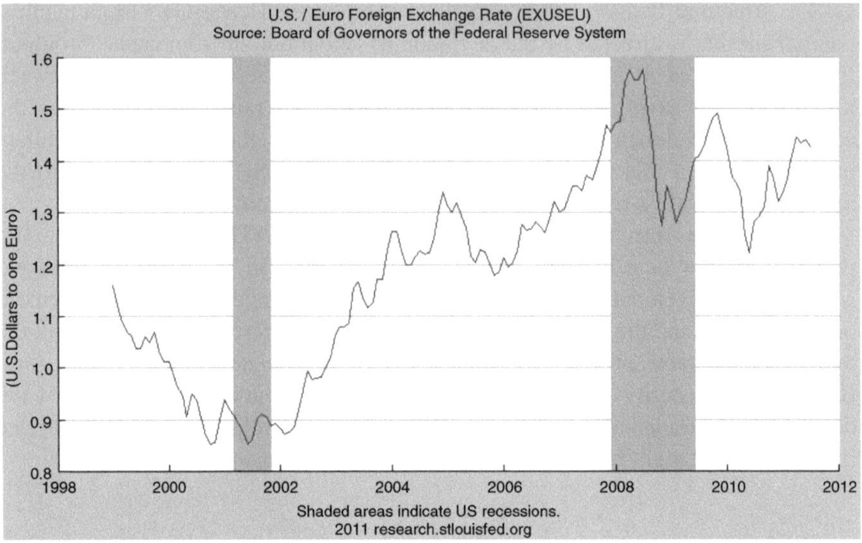

Fig. 8.5 Euro-$US exchange rate

References

Banco de Espagna. 2008. Annual report.

Bank for International Settlements, Eiglsperger, Martin, and Wim Haine, EU housing statistics IFC Bulletin #31 Annex 1. 2009.

Bank of Greece. Annual report.

Cabral, R. 2010. VOX The PIGS' external debt problem.

EEAG. 2011. *EEAG report on the European economy*. Munich: CESifo.

Fischer, Christoph. 2010. What can EMU countries' sovereign bond spreads tell us about market perceptions of default probabilities during the recent financial crisis? Niko Dötz Discussion Paper Series 1: Economic Studies No. 11/2010.

International Monetary Fund, IEO. 2011. IMF performance in the run-up to the financial and economic crisis.

International Monetary Fund. World Economic Outlook, Advanced economies: General government structural balances, cyclically adjusted.

Irish Banking Crisis. 2010. Regulatory and financial stability policy 2003–2008, Report to Minister of Finance by Governor of the Central Bank.

Moody's US Municipal Bond Rating Scale 2002.

Plagnol, Valerie. 2010. *European bond markets in financial crisis*. Reuters: Ecowin.

Stein, Jerome L. 2006. *Stochastic optimal control, international finance and debt crises*. Oxford: Oxford University Press.

Williamson, John (ed.). 1994. *Estimating equilibrium exchange rates*. Washington, DC: Institute for International Economics.

Index

J.L. Stein, *Stochastic Optimal Control and the U.S. Financial Debt Crisis*,
DOI 10.1007/978-1-4614-3079-7, © Springer Science+Business Media New York 2012

MIX
Papier aus verantwortungsvollen Quellen
Paper from responsible sources
FSC® C105338

If you have any concerns about our products,
you can contact us on
ProductSafety@springernature.com

In case Publisher is established outside the EU,
the EU authorized representative is:
Springer Nature Customer Service Center GmbH
Europaplatz 3, 69115 Heidelberg, Germany

Printed by Libri Plureos GmbH
in Hamburg, Germany